UiPath Administration and Support Guide

Learn industry-standard practices for UiPath program support and administration activities

Arun Kumar Asokan

BIRMINGHAM—MUMBAI

UiPath Administration and Support Guide

Group Product Manager: Alok Dhuri
Publishing Product Manager: Shweta Bairoliya
Senior Editor: Ruvika Rao
Technical Editor: Pradeep Sahu
Copy Editor: Safis Editing
Project Coordinator: Manisha Singh
Proofreader: Safis Editing
Indexer: Sejal Dsilva
Production Designer: Vijay Kamble
Marketing Coordinators: Deepak Kumar and Rayyan Khan

First published: September 2022
Production reference: 1250822

Published by Packt Publishing Ltd.
Livery Place
35 Livery Street
Birmingham
B3 2PB, UK.

ISBN 978-1-80323-908-8

www.packt.com

This book has been in progress for a long time, and I would like to thank my loving wife, Harini, for her relentless support during this tiring endeavor, and my children, Nakshatra and Yugan, for motivating me to be a role model. I would also like to thank my parents, Mr. Asokan (late) and Mrs. Selvi, for their inspiration to do great things and keep moving forward toward a noble target, come what may!

– Arun Kumar Asokan

Contributors

About the author

Arun Kumar Asokan is an Intelligent Automation enthusiast, educator, and leader who has experience setting up and managing **Automation Centers of Excellence (CoEs).** He has also authored the bestselling *Robotic Process Automation Projects* book. He has previously worked for different digital consulting groups in advocating process automation benefits and helping clients in different parts of the world including the US, Europe, Australia, and India.

He is based in the **Dallas Fort Worth** (**DFW**) area in the US and he holds a Bachelor of Technology degree from his hometown college, Pondicherry Engineering College, India, and an MBA from Queensland University of Technology, Australia.

I want to thank the people who have been close to me and supported me,
especially my wife, Harini, and my parents, Mr. Asokan (late) and Mrs. Selvi.

To all the RPA content creators and enthusiasts who motivated me to
write this content and the team at Packt for their help and support throughout
the process.

About the reviewers

Shaun Dawson has over 25 years of experience in the development and leadership of technology teams. Shaun was a co-founder of Virtual Operations, a leader in RPA development, deployment, and management services, and a very early UiPath partner. He went on to join Cognizant in 2014 to build out the BPS Intelligent Automation Practice, then served as the North American CEO of Robiquity, Inc.

Shaun has been responsible for the design and operation of several processes within automation, including assessment training, line of business onboarding, and the conception and build-out of automation CoEs for several UiPath clients. Today, he leads UiPath's Consulting Delivery organization. He lives and works in San Diego, CA.

Roopak Desai has over 20 years of broad IT experience with expertise in capturing and analyzing complex business needs and translating them into software design strategies, user experiences, and solutions. He is a highly analytical, dedicated, and results-oriented technical professional with a client-centric solution delivery approach. His experience includes leading automation platform services/support as part of the robotic process automation CoE in a reputed financial firm.

Table of Contents

Part 2: UiPath Administration, Support, DevOps, and Monitoring in Action

4

5

Robot Management and Common Support Activities   121

6

DevOps in UiPath   145

7

Monitoring and Reporting in UiPath   177

Part 3: UiPath Maintenance and Future Trends

8

UiPath Maintenance and Upgrade   205

9

UiPath Support – Advanced Topics and Future Trends   229

Preface

Hello there! **Robotic Process Automation (RPA)** is the practice of automating manual, repetitive, and time-consuming tasks performed by humans. This is achieved by completing those tasks using software programs called robots (bots). UiPath is a leading RPA software platform, and it strives to complement the core RPA capabilities with additional features to move enterprises toward an intelligent automation paradigm.

There are three main pillars for any RPA engagement:

- RPA strategy, governance, and resource management
- RPA delivery and program management
- RPA operation, service, and maintenance

There are many resources that cover the first two pillars, so this book aims to cover the third pillar. This book will present all the best practices for administering, monitoring, reporting, maintaining, and managing DevOps for your UiPath RPA program.

I will provide relevant information and guide you through some of the UiPath administration and support activities that happen in real UiPath programs based on two main sources of information:

- My experience and knowledge from the past few years in this process automation domain
- Interviews with UiPath RPA champions leading the UiPath CoE (at different stages of maturity) in different industries across the globe

As per the Analyst Report (Jan 2022), the UiPath platform is the leader in the RPA space. The product is expected to grow much faster, along with the proportional growth in the RPA adoption rate by companies, in the next few years. As the adoption rate and the maturity of the RPA programs increase, the mandatory UiPath administration, support, maintenance, monitoring, and deployment activities will become more challenging compared to developing the bots. This is one of the major hurdles for many firms that are looking to scale their UiPath RPA program.

There is a lot of demand for UiPath Support professionals and they have faster growth opportunity compared to UiPath development.

Who this book is for

UiPath RPA leads, developers, or any IT support person can gain practical insights into how to perform UiPath support and administration tasks from this book.

The three main personas who are the target audience of this content are as follows:

- RPA CoE leads: Leaders who are looking to set up or improve their UiPath support organization.

- RPA developers: They will learn all aspects of UiPath support and administration to add value to their current UiPath developer role. This book will help them grow in their career to be a lead, manager, and so on.

- Support personal: They will gain an overview of not just how to support the main components, such as orchestrators and robots, but also get a 360-degree perspective of the UiPath support and administration role. This content will help them succeed in UiPath support/administrator interviews as well. They will learn how to use industry best practices and tips and tricks from real life to perform efficiently in their work.

What this book covers

Chapter 1, Understanding UiPath Platform Constructs and Setup, provides an introduction to UiPath platform products and also explains the core architecture components. It will also cover the basic setup of Orchestrator, program organization, and robots.

Chapter 2, Defining UiPath Support Strategy, Framework, and Models, provides an overview of the UiPath RPA support framework and model. It shares the best practices during onboarding a new process in support and also covers alignment with the RPA **Center of Excellence (CoE)**, Enterprise monitoring, security, infrastructure, and teams. This chapter will also explain how and when internal and external teams including the UiPath support team need to be involved.

Chapter 3, Setting Up UiPath Support Enablers, provides an overview of UiPath RPA support components. In addition to this, you will learn how to set up RPA monitoring, reporting, and deployment. This chapter also covers how to set up continuous improvement with the help of alerts, and also touches upon custom common support utilities that will enhance RPA support performance.

Chapter 4, UiPath Orchestrator Administration, provides an overview of UiPath Orchestrator administration. You will learn how to use UiPath Orchestrator to perform system and tenant administration. The chapter will deep-dive into robot, license, process, and job management. Finally, other supporting features of Orchestrator will also be covered.

Chapter 5, Robot Management and Common Support Activities, provides an overview of UiPath robot administration and discusses real-life support activities. It then introduces the different priorities of support requests based on SLA and also covers how to handle complex support requests such as infrastructure and application migration.

Chapter 6, DevOps in UiPath, provides an overview of using DevOps concepts in UiPath programs. In addition to providing an overview of the overall automated delivery process, this chapter also introduces continuous integration and delivery using the UiPath Jenkins plugin and integration with GitHub. Finally, this chapter enables the user to understand how the change management process is integrated with the Jenkins pipeline.

Chapter 7, Monitoring and Reporting in UiPath, provides an overview of the UiPath monitoring and reporting framework. The next sections cover different types of monitoring and reporting at various levels of the framework, such as business, application, and infrastructure. This chapter finally covers different monitoring and reporting options.

Chapter 8, UiPath Maintenance and Upgrade, provides an overview of UiPath RPA platform maintenance and upgrade activities. It starts with regular platform maintenance activities such as database maintenance, and then touches on RPA infrastructure and supports platform maintenance as well. This chapter also covers the upgrade runbook, which will be really helpful for the UiPath support team.

Chapter 9, UiPath Support – Advanced Topics and Future Trends, provides an overview of UiPath RPA advanced support areas, such as the UiPath self-service catalog. Then it will introduce how different custom UiPath support and monitoring utilities can be built and utilized to add value to the setup.

The next section will deal with how support personnel will support UiPath requests related to IT security, risk, and audit. It also covers how to extend the core support principles to support the extension of UiPath RPA platform components such as the test suite, chatbots, document understanding, apps, data, and integration services as well. Finally, this chapter will cover future trends in the UiPath support space, such as automated support, containerized deployments, and multi-vendor ecosystems.

To get the most out of this book

You will need to have an understanding of the basics of the UiPath platforms and of IT application administration, support, and monitoring.

Software/hardware covered in the book	Operating system requirements
UiPath	Windows
Jenkins	
Amazon Web Services (AWS)	

Prerequisite steps

Before we begin with Chapter 1, we need to set up a few things. You can check the prerequisite steps for this book from GitHub at `https://github.com/PacktPublishing/UiPath-Administration-and-Support-Guide`. If there's an update to the steps, it will be updated in the GitHub repository.

We also have other code bundles from our rich catalog of books and videos available at `https://github.com/PacktPublishing/`. Check them out!

Download the color images

We also provide a PDF file that has color images of the screenshots and diagrams used in this book. You can download it here: `https://packt.link/mNCOr`.

Conventions used

There are a number of text conventions used throughout this book.

`Code in text`: Indicates code words in text, database table names, folder names, filenames, file extensions, pathnames, dummy URLs, user input, and Twitter handles. Here is an example: "Delete the `C:\~\SysWOW64\config\systemprofile\AppData\Local\UiPath\Logs\execution_log_data` folder."

A block of code is set as follows:

```
pipeline {
        agent any
        environment {
            ORCHESTRATOR_URL = https://cloud.uipath.com/~/
orchestrator_/"
            ORCHESTRATOR_LOGICAL_NAME = "XXXX"
            ORCHESTRATOR_TENANT_NAME = "YYYY"
            ORCHESTRATOR_FOLDER_NAME = "ZZZZ"
        }
```

Bold: Indicates a new term, an important word, or words that you see onscreen. For instance, words in menus or dialog boxes appear in **bold**. Here is an example: "A list of all available robots is accessed from the **Tenant | Robots** tab."

> **Tips or Important Notes**
> Appear like this.

Get in touch

Feedback from our readers is always welcome.

General feedback: If you have questions about any aspect of this book, email us at customercare@packtpub.com and mention the book title in the subject of your message.

Errata: Although we have taken every care to ensure the accuracy of our content, mistakes do happen. If you have found a mistake in this book, we would be grateful if you would report this to us. Please visit www.packtpub.com/support/errata and fill in the form.

Piracy: If you come across any illegal copies of our works in any form on the internet, we would be grateful if you would provide us with the location address or website name. Please contact us at copyright@packt.com with a link to the material.

If you are interested in becoming an author: If there is a topic that you have expertise in and you are interested in either writing or contributing to a book, please visit authors.packtpub.com.

Share Your Thoughts

Once you've read *UiPath Administration and Support Guide*, we'd love to hear your thoughts! Scan the QR code below to go straight to the Amazon review page for this book and share your feedback.

https://packt.link/r/1-803-23908-5

Your review is important to us and the tech community and will help us make sure we're delivering excellent quality content.

Part 1:
UiPath Platform and
Support Setup

In this part, you will get an overview of the UiPath platform and learn about the UiPath Orchestrator and robot setup. In addition to this, you will also learn about the RPA support framework and model, the best practices in designing and implementing an RPA support strategy, RPA support components, and finally, the best practices in setting up deployment, monitoring, reporting, and continuous improvement initiatives.

This section contains the following chapters:

- *Chapter 1, Understanding UiPath Platform Constructs and Setup*
- *Chapter 2, Defining UiPath Support Strategy, Framework, and Models*
- *Chapter 3, Setting Up UiPath Support Enablers*

1

Understanding UiPath Platform Constructs and Setup

Robotic process automation (**RPA**) adoption is expected to grow at a much faster rate in upcoming years. As UiPath Platform is one of the market leaders in this space, the demand for UiPath Platform will grow, and so will the importance of UiPath support and administrator roles in the companies. Let's get started with the first chapter.

The first chapter will introduce UiPath Platform products and explain the core Orchestrator architecture components. It will also cover the basic setup of Orchestrator, Robots, and the UiPath program organization.

Here is what you will learn as part of this first chapter:

- Get an overview of UiPath Platform
- Understand UiPath Orchestrator and Robots setups
- Difference between on-prem and cloud setup
- Learn about Robot types and licenses in detail
- Real-life environment setup

UiPath Platform overview

UiPath is an RPA platform company that was started back in 2005. It began to get real traction in 2015, and as of 2022, it is the leader in the RPA vendor landscape. The core product lineup is made up of the three products listed here, as any basic RPA use cases can be automated, deployed, and monitored with these core products:

- **UiPath Studio**: The developer platform to create the automation scripts
- **UiPath Orchestrator**: The core platform to run and monitor the Robot's jobs
- **UiPath Assistant**: Robot client software installed in runtime environments to execute the jobs

All UiPath customers need to set up this three-core software to start their automation journey. As UiPath evolved into a hyper-automation platform, it started to add new products to complement its core process automation offering. It's important for any UiPath support and administrator to have high-level knowledge of all the UiPath products. They might need to set up and support them as and when required.

> **Note**
>
> The products can be installed separately on-prem and subscribed as a cloud package in the UiPath automation cloud. Recently, the Automation Suite offering combines all these different products and can be leveraged as a Kubernetes container deployed and used on demand.

There is a whole suite of UiPath products to support the entire automation journey of any organization. The product inclusion is solely dependent on the UiPath Enterprise customer needs, and the list of available choices is listed in the following figure:

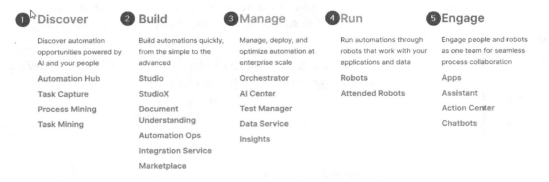

Figure 1.1 – UiPath product portfolio (as of January 2022)

> **Note**
>
> Reference information for all the UiPath products can be found at this link: `https://docs.uipath.com`.
>
> Starting from the 2021.10 LTS release, the UiPath suite supports Linux-based OS, and the new offering, Automation Suite, is preconfigured with Kubernetes containers and management tools.

UiPath Enterprise customers can choose cloud or on-prem versions of the products based on their needs.

In a few cases, they might even switch from an on-prem version of the products to a cloud version or the other way around. As all the products are supported in both editions, the UiPath support and administration team must be educated and trained to handle both scenarios.

Suppose a new product, say process mining or test manager, is added to the UiPath environment. In that case, the UiPath support and administration team needs to be equipped to start supporting these additional products along with the core software, such as Orchestrator, Studio, and Robots/Assistant.

Let's consider two UiPath support administrators named as follows:

- Patty, who works at XYZ Bank, needs to set up Orchestrator on-prem
- Candace, who works at ABC Insurance Corporation, needs to enable the cloud Orchestrator in her organization to explain the concepts in the following sections

On-prem UiPath Orchestrator

Learning about UiPath Orchestrator's inner details is mandatory for any successful UiPath support and administration professional. If Orchestrator is installed on-prem, it is even more important to learn about it. There are two ways Orchestrator can be configured:

- **Single node setup** – One instance of UiPath Orchestrator is installed
- **Multi-node setup** – Multiple instances of UiPath Orchestrator are installed

Before Patty can start the actual installation, she must understand the underlying architecture of these two options.

Let's see them in detail in the next section.

Single node UiPath Orchestrator architecture

As the name suggests, a single node will have Orchestrator installed in one instance. As per UiPath documentation, a single node can support small or medium-scale deployment of Robots (i.e., 1 - 250 unattended Robots or 1 - 2,500 attended Robots). There is no failover plan, hence downtime is expected during maintenance windows or server outages.

This is the preferred model for any proof of concept and pilot run since the time and cost to set this up are low:

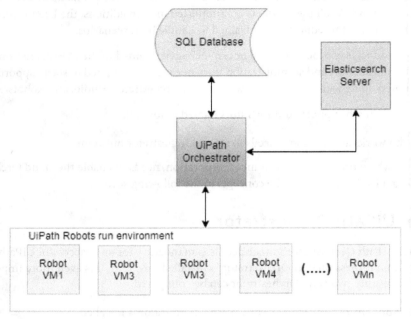

Figure 1.2 – UiPath single node architecture

Components list

- **UiPath Orchestrator**: It is recommended to install the Orchestrator application on a Windows server (physical or virtual). It will be a web application accessible from the **Internet Information Services (IIS)**.

- **SQL database (DB)**: A SQL database is recommended to be set up in a separate SQL Server, or you can even leverage the cloud services provided by popular cloud service providers. This database will be the primary application database where the data can be queried later after installation. The connection must always be alive between the DB and UiPath Orchestrator.

- **UiPath Robots (digital workers)**: This software is set up in virtual or physical machines that can be remotely accessed by Robot accounts. Multiple devices can be registered and connected with Orchestrator to execute the automation.

- **Elasticsearch (ES) and Kibana (optional)**: Orchestrator logs can be directed to the SQL DB and ES. Hence, ES is an optional setup. ES and Kibana can be set up on the same Orchestrator Windows server or a separate server based on the need. ES is an open source application that can optimize UiPath Orchestrator logs access from Orchestrator. We can redirect the execution logs away from default DB tables and into the ES indexes by updating the information in `UiPath.Orchestrator.dll.config`. Kibana is an optional open source monitoring dashboard application that can also be installed on this server for monitoring needs.

  ES and Kibana are part of Elastic Stack, and you can learn more about Elastic Stack here: `https://www.elastic.co/elastic-stack/features`.

> **Note**
> UiPath Insights is the official business analytics and monitoring tool offering from UiPath. It provides complete monitoring and alerting capabilities for a UiPath program. This product can also be installed when we set up the Orchestrator nodes. If Insights is in use then ES does not need to be set up.

Patty has gathered enough knowledge to set up a single node Orchestrator now.

> **Note**
> Please refer to the software requirements to install the supported version of the software from here: `https://docs.uipath.com/installation-and-upgrade/docs/orchestrator-software-requirements`.

Multi-node UiPath Orchestrator architecture

As the name suggests, multi-node will have multiple instances of Orchestrator installed. UiPath documentation can support large-scale UiPath Robot deployments with better performance and fault tolerance, as multiple Orchestrator nodes are available. If one node fails, other nodes will process the transactions and bots.

In addition, horizontal scalability is possible as we can add additional nodes to the Orchestrator as the Robot needs increase. Starting with two Orchestrator nodes that support up to 7,000 Robots, we can scale to 15 Orchestrator nodes, where we can have 150,000 Robots on the platform.

Patty and her team can start building a single Orchestrator instance. Then, as they harden things up for production, they can move ES off the single server, add additional DB server ES shards, and bring up a second node as an active/standby configuration. In other words, single to multi-node is more of a progression for any organization who are willing to mature their automation journey.

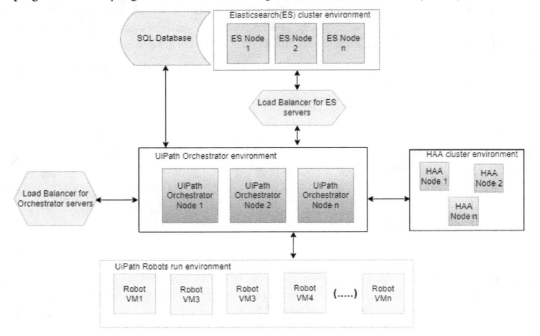

Figure 1.3 – UiPath multi-node architecture

Let's look at the components one by one:

- **UiPath Orchestrator scale set**: It is recommended to install the individual Orchestrator applications on a dedicated Windows server(s). All the Orchestrator nodes should have a similar IIS configuration, access rights, and security policies. It is recommended to have a service windows user account configured to this UiPath Orchestrator scale set. This can be used to set up the DB connection.

- **Scalable SQL DB**: The database server should be configured based on the number of Orchestrator nodes in production. Most cloud DBs are now scalable, and it would be good to consult the DB administrator to set up the best practices and proper access privileges.

  DB maintenance activities need to be put in place from day one because the performance of Orchestrator is directly related to DB health.

- **UiPath Robots (digital workers) run environment**: These are set up in virtual or physical machines, which can be remotely accessed by Robot accounts. The only difference here is that this environment should be set up to scale faster to accommodate the new Robots quickly. A Kubernetes container is a popular option recently in this environment.

- **ES cluster environment**: It is recommended to set up a separate Windows server and host the ES and Kibana applications to build this cluster. The main idea is to have a scalable option when the Robots' logs start to increase with the Robots' jobs being executed. Logs are critical to maintaining UiPath operational health, and hence maintaining this cluster is crucial to support and monitor.

- **High Availability Add-on (HAA) cluster environment**: HA provides redundancy and stability for your multi-node Orchestrator deployment through failure resistance. In an HAA configuration, as long as multiple Orchestrator and HAA nodes are available, the other nodes will be activated if one fails; if any nodes fail or are taken down on purpose, processing will "failover" to the rest of the nodes in the cluster. Moreover, horizontal scalability is also possible that means you can add another node whenever your Robot counts needs to grow. UiPath HAA is just Redis **Original Equipment Manufacturer (OEM)** version for UiPath. Redis is used for in-memory data structure storing used for caching, which will improve the application's performance. The UiPath support contract supports only the multi-node setup with UiPath HAA. Hence, it is more than a nice-to-have component. Although there are other ways to set up multi-node failover and horizontal scaling, UiPath HAA is the only method supported by UiPath.

  Please refer to the documentation for setup: `https://docs.uipath.com/installation-and-upgrade/docs/haa-installation`.

- **Load balancer (LB)**: A load balancer is an appliance (which encompasses hardware load balancers, such as F5, as well as software load balancers) that will automatically distribute incoming web traffic across various endpoints. All the cloud service providers have a load balancer offering that can be leveraged, for example, F5 load balancing on AWS. UiPath Orchestrator supports Layer 4 load balancing only, not Layer 7, and no SSL offloading:

  - **LB for Orchestrator servers**: Load balancers are critical to redirect traffic to different Orchestrator nodes originating from various client requests. It is recommended to have a load balancer URL for Orchestrator login so if a node is down, we can still access the application.

  - **LB for ES servers**: The same concept applies to redirect connections from Orchestrator to ES servers. If there are multiple ES servers, then a load balancer is functional. There can also be scenarios where there are numerous ES shards on several different servers without a load balancer.

- **UiPath Identity server:** This provides a centralized authentication service to all UiPath products, and it is included in the on-prem Orchestrator installer. It is recommended that the enterprise **Identity and Access Management (IAM)** team configure the Identity server component to meet the existing enterprise standards. Most of the configurations are done on `appsettings.json` (please refer to `https://docs.uipath.com/installation-and-upgrade/docs/orchestrator-is-appsettings-json`).

The recommended hardware setting for scaling up the operation is mentioned here: `https://docs.uipath.com/installation-and-upgrade/docs/orchestrator-hardware-requirements`.

> **Note**
>
> NuGet packages are stored in a shared location that all the Orchestrator nodes can access. SQL Server will have redundancy and have the **Always-on Availability group**. If UiPath Insights is used, it is good to have an Insights application server added to this architecture.
>
> Please refer to this link for hardware requirements based on the number of Robots in production: `https://docs.uipath.com/installation-and-upgrade/docs/orchestrator-hardware-requirements`.

I hope this was interesting. Patty and her team know to set up a multi-node Orchestrator in their organization. Now, let's look at the Orchestrator setup details in the next section.

Orchestrator setup

As introduced in the previous section, UiPath Orchestrator is the critical component used for the creation, monitoring, and deployment of Robots in the defined runtime environments.

There are three main options of UiPath Orchestrator setup possible in an enterprise:

- The first one is a standalone installation which can be on-prem or cloud
- The second one is the UiPath Automation Cloud
- The third option is to set up the UiPath Automation Suite (starting November 2021)

It is important for UiPath Support and Administrator professionals to overview the UiPath Orchestrator setup. Let's start with this setup of a single node on-prem Orchestrator.

On-prem UiPath Orchestrator (single node)

Patty can now proceed with the installation detailed in this section. On a high level, the UiPath Orchestrator setup can be charted down into five different steps, as shown in the following diagram:

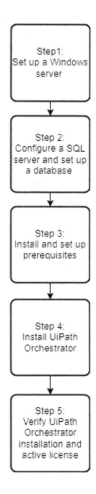

Figure 1.4 – UiPath single node setup steps

All five steps are discussed in this section; lets start with the first step:

1. Set up a Windows Server on a machine or a cloud infrastructure. Ensure this web application server configuration (CPU, memory and security groups, and so on) is set up as per the recommendation from the UiPath documentation.

    Refer to this link for more information: https://docs.uipath.com/installation-and-upgrade/docs/orchestrator-hardware-requirements.

2. Configure a SQL server and set up a DB. The next important step is to set up the DB server. Again, this can be set up on a separate machine or subscribed as a cloud service. Once the server is set up, ensure the DB is created and credentials noted.

3. Install and set up the prerequisites. The prerequisites (URL Rewrite, .NET, certificates, and so on) need to be installed on the application server before starting the UiPath Orchestrator setup. Please refer to the complete checklist from this link: `https://docs.uipath.com/installation-and-upgrade/docs/orchestrator-prerequisites-for-installation`.

4. Install UiPath Orchestrator; this is the primary step of running the UiPath Orchestrator installer and updating the mandatory information such as the DB information, Identity server, certificate, and so on). The installation will be completed successfully when all the prerequisites and access are correctly configured. Otherwise, the installer will roll back the installation.

5. Verify the UiPath Orchestrator installation and active license. The last step is to verify the installation by logging into Orchestrator and activating the license.

> **Note**
>
> The ES server setup mentioned in the architecture diagram is optional, and we will cover it in a different chapter. Please refer to *Setting up On-Prem UiPath Orchestrator in AWS EC2* in the book's GitHub repository (`https://github.com/PacktPublishing/UiPath-Administration-and-Support-Guide`) to get a step-by-step walkthrough of this setup.

Next, let's look at some of the essential considerations for a multi-node setup for Patty and her team.

Multi-node on-prem Orchestrator considerations

Based on the customer's need, a multi-node setup can be made. In the case of Patty and her team, once a single node orchestrator was set up, they decided to add more nodes and grow into a multi-node resilient platform. Multi-node setup is the most common Orchestrator setup in any organization that is still using an on-prem environment. A two- or three-node setup is typical for mid to large-scale Robot setups. There are a few considerations to set up this environment:

- The Orchestrator setup needs to replicate to Orchestrator cluster setup. Multiple servers will act as nodes for UiPath Orchestrator. Single node setup can be scaled to multi-node with appropriate infrastructure if needed.

- Similarly, the ES setup needs to be scaled up to the ES cluster environment. Multiple ES servers may need to be set up to support the scaling needs of Robots.

- SQL servers need to be configured to be scalable and available. It is recommended to have a redundancy setup in effect for the servers.

- In-memory HAA clusters need to be configured in the multi-node setup to maintain the platform's performance.

- Load balancers need to be configured to redirect traffic to Orchestrator, and an additional load balancer is recommended if there is an ES cluster setup.

- The Robots farm needs to be set up to scale up operations on demand.

The entire setup can be made on physical machines or cloud infrastructure based on companies' IT policies. After following the steps, Patty and her team set up multi-node Orchestrator.

Cloud – UiPath Orchestrator

In recent years, a cloud UiPath Orchestrator has been gaining traction, and as many enterprises move toward cloud-first architecture, let's cover this aspect.

UiPath will maintain the complete infrastructure for Orchestrator, and all the resources will be held in the cloud. As UiPath introduces new products regularly, it becomes easy for customers to quickly subscribe to them and not install them as separate applications in on-prem environments.

Let's consider ABC Insurance Corporation, an insurance service provider and a UiPath Enterprise customer who is setting up the cloud UiPath platform. Candace and her team at ABC Insurance Corporation will work on this assignment.

Automation Cloud setup

Now it's Candace's turn to set up the cloud Orchestrator for her team. The UiPath account team will provide the initial cloud administrator for UiPath Automation Cloud. She can add new users from the admin and access the Orchestrator default tenant, and the rest of the setup can be made once we log in to cloud Orchestrator.

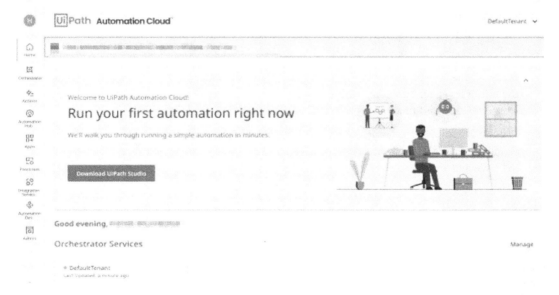

Figure 1.5 – UiPath Automation Cloud

This is the primary offering for all new customers. Many existing customers are interested in converting to the cloud as it reduces the overhead of maintaining the Orchestrator platform and easy scalability.

Even though there are plenty of advantages, there are also a few disadvantages, such as not having access to the DB. Custom monitoring with Kibana or any internal platform is also not possible in this setup.

Candace and her team could access the UiPath cloud Orchestrator from the Automation Cloud within minutes. Next, she and her team need to understand Robots and options provided by UiPath before setting up.

Robots setup

As per UiPath documentation, UiPath Robots is the client program that runs processes developed in UiPath Studio or Studio Pro. Robots information is stored and controlled by UiPath Orchestrator.

There are different ways Robots can be configured in an Enterprise setup based on business needs. Before we can configure Robots, we need virtual machines to host these Robots' client software.

Before moving forward, let's understand folder concepts, as Robot types are tied to folder concepts.

UiPath folders

Orchestrator resources, such as processes, jobs, and assets, can be logically grouped into folders. Folder concepts provide an excellent way to organize the entire UiPath program. There are two types of folders currently supported by UiPath:

- **Modern folder**: This is the recommended option that will enable the UiPath program to utilize better all the latest features available in the UiPath Platform. Users and roles are assigned at the folder level, and greater flexibility to move Robot licenses between folders. Automation workflow executed in a folder can access resources in another folder, too.

- **Classic folder**: User roles are assigned at the tenant level, and Robots can be only part of a single folder. If the Robot needs to be used in a different classic folder, then the Robot needs to be deleted and recreated in the new folder.

Note

Classic folders will not be supported from UiPath Orchestrator v2022; all the existing customers need to migrate to modern folders to upgrade to this version.

UiPath Robot machine setup

Machines are runtime environments (physical or virtual) where the Robots execute the process. There are four options of machines in the latest version of Orchestrator (v2022.10):

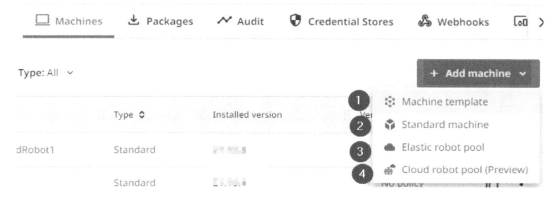

Figure 1.6 – Machine options

> **Note**
> The software needed to run the automation (including UiPath Assistant) must be installed once the machine is set up. Access to resources, such as network shared location, needs to be provided.

All the Robot machine options are discussed in this section. Lets look through the options:

- **Machine template**: Robot machines can be logically grouped as per an organization's functional or IT needs. Let's assume 10 AWS EC2 virtual machines are available to execute HR and supply chain processes. We can build two different templates for HR and the supply chain. The HR template might have Version 10 of App A and App B installed, and the supply chain process needs Version 12 of App A and App C. Best practice here is to create a "Gold Image" with all the applications you will need for your automation. Failing that, you need to plan out and create machine templates for the combinations of applications you need. Adding a machine template key to the Robots will break the machine mapping.

> **Note**
> As the Robot and machine mapping is broken, any program can be executed in any Robot tied to the machine template.

To build a template, fill in the template name and runtime licenses needed, for example, a **Production (Unattended)** 5 runtime license. Additional options are to mention whether the machines are Windows- or Linux-based once under the **Process Compatibility** settings:

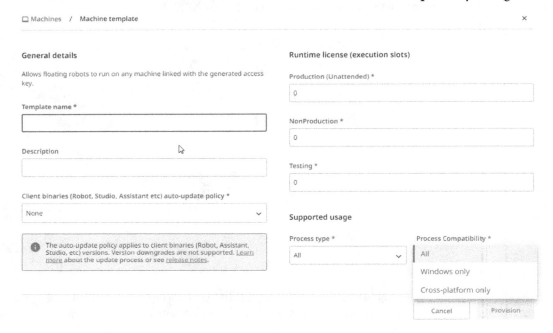

Figure 1.7 – Machine template options

In addition to this, we can also customize it to run foreground or background jobs.

- **Standard machine**: This is the default option available, and runtime licenses are needed. Most of the classic folder Robots are running on standard machines, and it is one of the most popular machine types used currently in organizations. The standard machine is supported in the modern folders concept, too. This will be a recommended type with a robust bot machine mapping, for example, a dedicated bot running on email processing on a machine.

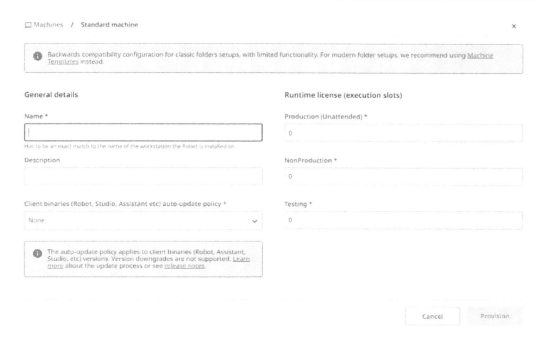

Figure 1.8 – Standard machine options

- **Primary Robot Pool**: This is one of the recent machine types added at the start of 2021. This is a scalable auto option provided by popular cloud vendors, which will allow to automatically add new Robot machines in the cloud as per the configured parameter such as cost, maximum number, and time.

We need to add a cloud provider connection before using this feature.

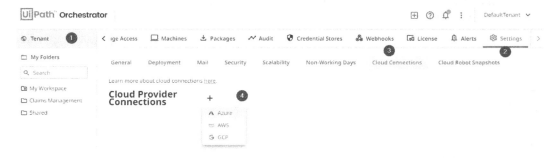

Figure 1.9 – Cloud machine options

We just need to provide a name and associate the cloud connection to enable these auto-scalable machines:

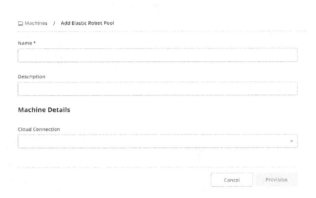

Figure 1.10 – Robot pool

> **Note**
>
> A Robot pool can be a way to reduce your license and infrastructure costs significantly. If your organization has a mature cloud practice or is trying to start one, it is highly recommended to consider using a Robot pool as part of your architecture plan. Please refer to this documentation to set up the cloud connection: `https://docs.uipath.com/orchestrator/v0/docs/elastic-robot-orchestration`.

- **Cloud Robot Pool (Preview)**: The last one is the latest feature in preview mode as of Jan 2022. This is a pure-play attempt by UiPath to enable Robots as a service offering. The Robot VMs will also be managed by the UiPath cloud platform and can be provisioned at will if this option is enabled. Reduction in risk and cost of maintenance are significant benefits in this approach, but lack of control, customization options, and Azure cloud dependency are a few drawbacks to be highlighted.

A new license type called **Automation Cloud Robots** is needed to leverage this feature. The user needs to enter a name and choose a machine image to set up a VM on UiPath Cloud. Once the virtual machine is enabled, the remote desktop needs to be enabled, and when we log in, UiPath Studio and Assistant are preconfigured:

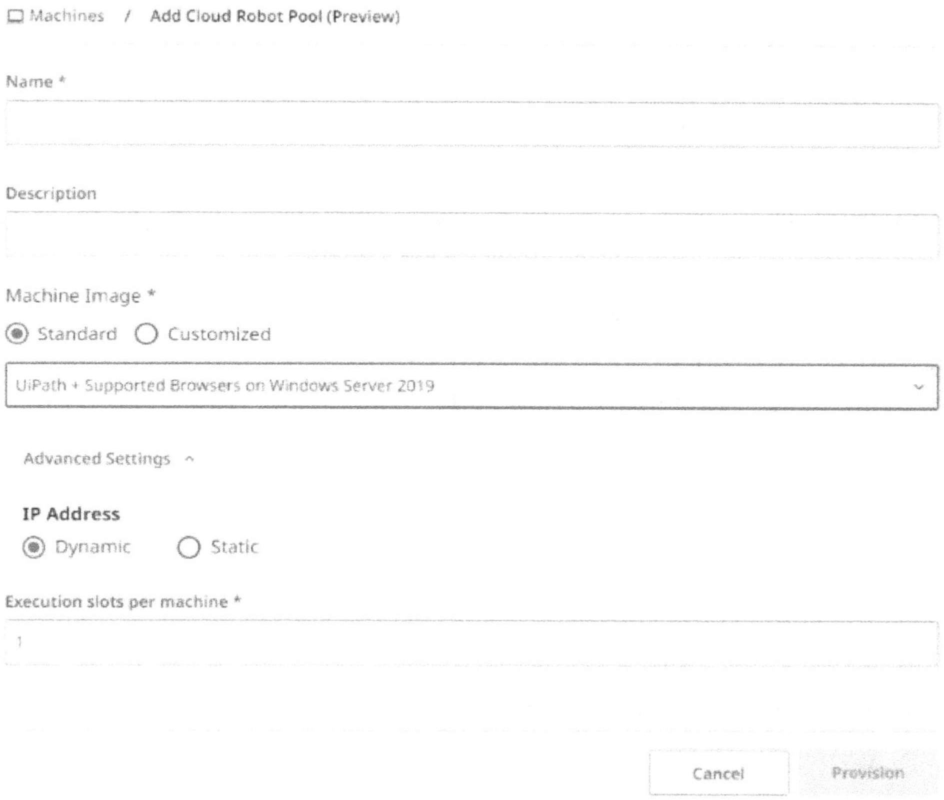

Figure 1.11 – UiPath Cloud Robot Pool

> **Note**
>
> I predict that this will be the machine of choice in a few years as UiPath will handle the VM maintenance, and clients have one less maintenance function. Please refer to the official documentation here: https://docs.uipath.com/orchestrator/v0/docs/automation-cloud-robots-vm.

UiPath Robot categories

Now that we have seen different machine options, let's look at some Robot categories.

Supervised versus autonomous Robot

The most common differentiation of Robots is based on the human-in loop concept, where the Robot executes a job with or without assistance from humans.

Supervised Robots run under human supervision and need humans to interact to complete a transaction:

- **Attended** – Bots operate along with the user and can be invoked via user events
- **RPA developer bots** – This is used to connect your Studio/StudioX or StudioPro to UiPath Orchestrator, and it is used when developing the UiPath scripts

On the other hand, autonomous Robots do not require human supervision to execute jobs:

- **Unattended** – Unattended bots are executed as batch jobs in the background, typically in a virtual environment, and are used to automate processes designed by the developers. It is the most popular bot license type being used in many organizations.
- **Non-production/testing bots** – Works in unattended/test mode for development purposes only.

> **Note**
> Choosing the right mix of Robot licenses is also an essential consideration for the UiPath program to succeed.

Standard versus floating Robot license

A standard Robot is configured in a dedicated virtual or physical environment, that is, there is a tight bot-to-machine mapping, and the same bot cannot be used in a different machine at the same time. The standard machine is the only machine type that works with a standard Robot. Standard Robots are the current standard in most organizations.

The floating Robots concept allows the Robot to be in multiple virtual or physical environments. The Robot is not associated with any machine. If a machine template is used and the Robot is configured to run on it with multiple machines, it becomes a floating Robot.

> **Note**
>
> Once the classic folder concept is depreciated in UiPath Orchestrator v2022, all the Robot licenses will be floating in nature by default on modern folders.

Normal versus high-density Robots

The subsequent differentiation of Robots is based on the hosting machine. If the Robot is hosted in individual VMs or desktops, it is considered a normal Robot.

If multiple Robot IDs are configured to run on a server environment, for example, Windows Server 2019 with five Robots accounts, it is a **high-density Robot**. The main advantage is that we can run multiple automation simultaneously in separate user sessions on the same machine. For this, some configurations must be done on the Windows Server machine first; then, you need to set up the environment in Orchestrator. Please refer to the following documentation: `https://docs.uipath.com/installation-and-upgrade/docs/setting-up-windows-server-for-high-density-robots`.

There are a few advantages and disadvantages of high-density Robots. One of the main advantages is to reduce the Robot machine maintenance downtime and easy monitoring of machine health. Still, the major drawback is if the server goes down with an issue, then multiple bot accounts will be stopped. One way to mitigate this risk is to have a machine template and provide additional runtime licenses to the machines in the template.

There are a few more types of Robots worth mentioning:

- **Test Robot**: If the UiPath Test Suite is enabled, we can procure this test non-production license type with a testing runtime associated. Converts a non-production Robot to a testing Robot.

- **AI Robot**: This is a particular type of license for the AI Center. An AI Robot license is used for **machine learning** (**ML**) training jobs, and a single license can also be used for executing any two ML skills available in the AI Center.

This section has covered all the major Robot types, and next, let's see how to set up a Robot. Candace and the team are ready to set up unattended Robots using the machine template option in the ABC Insurance Corporation's cloud Orchestrator.

UiPath Robot setup

On a very high level, both attended and unattended Robots can be set up on any of the four supported machine types, and they follow a structured flow, as depicted in the following diagram:

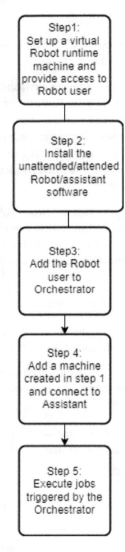

Figure 1.12 – Robot setup steps

Details of all the steps depicted in the diagram are detailed in this section, lets get into the details of the first step.

1. Set up a VM and install all the required software for executing the process as needed, such as an Office suite, SAP client, and so on. We must add the root user to the right user group and domain that has access to this machine. Usually, the organization's infrastructure group will be responsible for setting up the machines as needed, as per the policy agreed with the Enterprise architecture group.

2. Install the UiPath Assistant software and choose whether you wish to run attended or unattended automation on this machine. The Enterprise support team will be able to install the software with elevated privileges.

3. Next, we need to add a Robot user in Orchestrator with runtime licenses configured and provide correct access privileges to the folders and their resources.

4. Establishing the connection between Robots and Orchestrator is mandatory, and again, the Enterprise support team needs to get involved as admin privileges are required to update the Orchestrator settings.

5. Once the connection is established, trigger a test job, and see the execution in action.

> **Note**
>
> Please refer to *Setting Up the Cloud Orchestrator and Robots* in the book's GitHub repository (`https://github.com/PacktPublishing/UiPath-Administration-and-Support-Guide`) for step-by-step instructions to connect an unattended Robot to Orchestrator. In the real world, once a new Robot is set up on a machine, prerequisite checks are needed, such as whether the Robot has all the access rights enabled, software access, and licenses provided, such as Office 365 Acrobat, and so on. A trial job run is highly recommended in an attended mode before the actual runs take effect.

This completes the information on Robots. Candace and the team have followed the previous steps to set up the Robots.

Now that we've set up the UiPath environment, let's talk about organizing the UiPath program.

Program organization

On a high level, there are different ways to organize the UiPath RPA program to build for scale. Various organizations deploy different strategies based on agreed policies.

In this section, let's talk about two strategies to define a UiPath program organization. Program sponsors and architects usually characterize them, but UiPath support and Administrator teams should also be consulted when designing the structure. The UiPath Administration team will be the responsible party for the actual set of this model in UiPath environments.

We will use the ABC Insurance Corporation's UiPath RPA program as a reference to explain the concept better.

Organizing tenants and folders

When we install UiPath Orchestrator to an on-prem or set up on the cloud, we will get a **Default** tenant visible when we log in for the first time. Every organization starts with one tenant for the proof of concept and pilot. Once they get requirements, they mature and organize the UiPath resources in a multi-tenant environment.

It is essential to understand the meaning of the terms and their significance:

- **Tenant**: This is the highest level of grouping UiPath Orchestrator resources such as Robots, processes, machines, and so on. It is always recommended to have multiple tenants, usually at the organization function level such as finance and HR. You can log in to Orchestrator and learn more by viewing how the context changes:

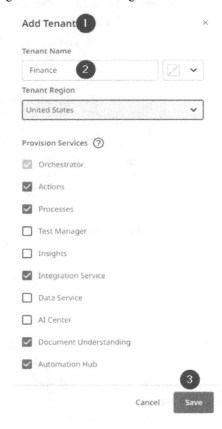

Figure 1.13 – Tenant setup

- **Folders**: This is the next level of grouping resources such as processes, assets, and so on, and it is usually at the business process level. The Robot's and user's rights can be defined at the individual folder level or inherited from the Level 1 folder. IT service management is a good example of a folder that will contain processes such as incident management, problem management, and so on, that is, UiPath Service Now integrations jobs. Modern folders are used in ABC Insurance Corporations' UiPath setup.

> **Note**
> We can add a subfolder as well to have more control at the subprocess level.

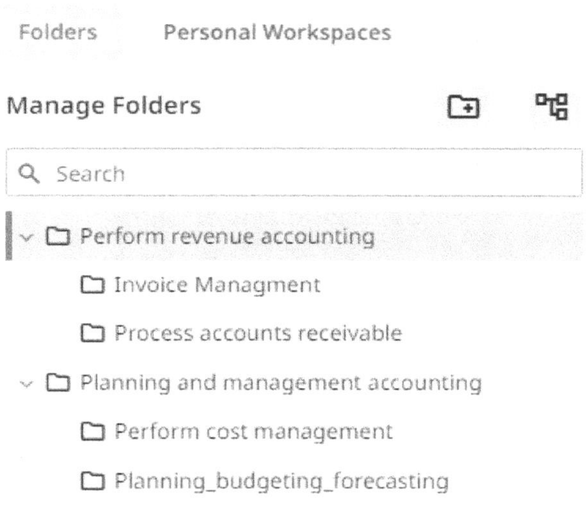

Folders   Personal Workspaces

Manage Folders

Q Search

∨ ☐ Perform revenue accounting

☐ Invoice Managment

☐ Process accounts receivable

∨ ☐ Planning and management accounting

☐ Perform cost management

☐ Planning_budgeting_forecasting

Figure 1.14 – Folders setup

> **Note**
>
> We have one more segregation called **Environment** supported in classic folder setups only, and it is very similar to the subfolder concept. We can group Robots and processes to dedicate them to a partial environment.

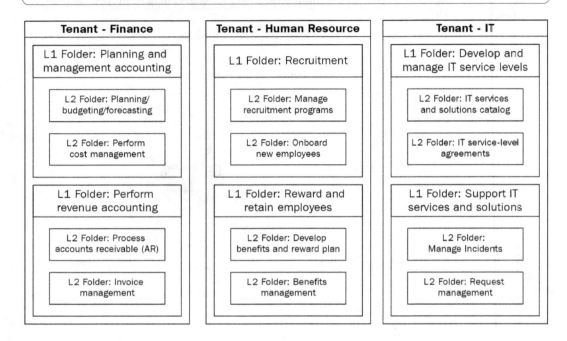

Figure 1.15 – ABC Insurance tenant/folder structure example

This was how the ABC Insurance UiPath program was organized, and there are many other approaches to manage the Robots, jobs, and other assets. The principal rule in many firms is that the team that pays for the license decides the strategy.

Organizing the RPA environments

The next level of organization is to replicate the above tenant/folder (along with all the resources) structure in different environments, namely test, **User Acceptance Testing** (**UAT**), and production.

The ABC Insurance Corporation's UiPath program setup is depicted in the following diagram; let's discuss the concept in detail:

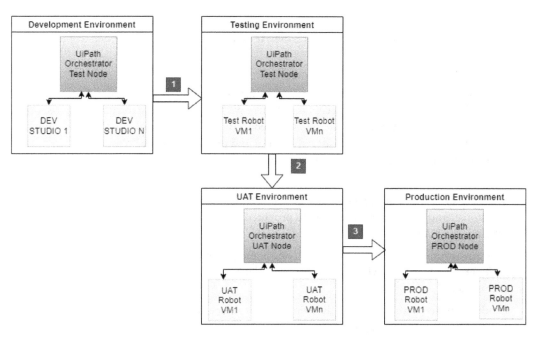

Figure 1.16 – ABC Insurance Corporation's UiPath environments

The details of the four environments depicted in the diagram are explained in the following sections, lets start with the development environment.

- **Development environment**: This is the primary environment where ABC Insurance Corporation's UiPath developers use Studio/Studio Pro or Studio X to create the automation. Ideally, individual developers will have separate licenses allocated to them. The Studios are usually connected to the UiPath test Orchestrator node. The developers also perform unit testing in this environment; then, the packages can be published in the test Orchestrator once the connection is established.

- **Testing environment**: This is the primary environment where ABC Insurance Corporations UiPath testers perform their system and integration tests. The UiPath test Orchestrator (non-production license type) and non-production Robot are usually configured in this environment. The tenant/folder setup and access rights for different UiPath resources need to sync in the test/UAT and production environment. The jobs that are created from the package from the Studio can be trigged in the test environment. Multiple test bots can be configured to perform the test as per the business need. The testers will usually use this environment to run functional, integration, performance, and security tests.

  The test environment access for the target application(s), such as Salesforce and Mainframe, needs to be provided to the bots in this UiPath environment to perform the tests.

> **Note**
> If you use the UiPath Test Automation Suite, then testing Robot licenses are needed to execute the test automation.

- **UAT environment**: This is the primary environment where ABC Insurance Corporation's UiPath business users and user acceptance testers perform UAT. Once the package passes the test gateway, the same package can be uploaded to the UAT environment. This environment again has a non-production license Orchestrator and non-production Robots configured. The setup should ideally reflect the production environment as the test will be equivalent to dry runs in production.

  The UAT environment access for the target application(s) that are automated such as SAP, Salesforce, and Mainframe, needs to be provided to the bots in this UiPath environment to perform the UAT tests.

  This environment will be used by the business analysts/product owner, who will provide the UAT signoff. Once the UAT signoff is provided, the package will be promoted to the production orchestrator.

- **Production environment**: This is the primary environment where ABC Insurance Corporation's UiPath business validators provide final validation and where the bots will execute the business transactions. This environment is the final runtime environment for the bots. This will have a production Orchestrator setup (single/multi-node), and production licensed Robots (unattended/ attended) on VMs or servers. The automation packages that business stakeholders signed off on will be run in this environment.

  The production environment access for the target application (automated, such as SAP, Salesforce, and Mainframe) needs to be provided to the bots executing the process in this UiPath production environment to perform the actual business transactions and processes.

> **Note**
> Based on the maturity of the UiPath programs, one or more of the environments may be missing in your current setup.

If the production is a multi-node Orchestrator setup, it is recommended that both test and UAT are multi-node setups. Risk reduction during Orchestrator upgrade is the primary reason behind this recommendation. The test and UAT UiPath Orchestrator nodes will be upgraded to a newer version before the production orchestrator is upgraded. The issues can be identified upfront and will improve the success rate of the production upgrade.

We mentioned the manual movement of the UiPath NuGet packages between environments by the ABC Insurance Corporation's RPA team. It doesn't always have to start that way, and the whole deployment process can be automated from the start of the program. Automated package movement through a DevOps pipeline has an extensive list of benefits to the UiPath RPA program. We will look at this UiPath DevOps concept in a later chapter.

Summary

UiPath RPA Platform is a robust process automation platform with many products in its portfolio. I hope you got a quick overview of the core UiPath products currently shipped as part of UiPath Platform and on-prem and cloud offerings. We used two personas from XYZ Bank and ABC Insurance Corporation in this chapter to explain the concepts so that you can relate to the idea of real-life personas.

In addition to this, you would have understood the on-prem and cloud architecture of UiPath Orchestrator. This is one of the most important concepts for any UiPath support and administrator role.

Next came the different types of Robot types and licenses offered. Having complete knowledge of Robots will give you an edge as a UiPath support professional, to provide valuable inputs during the UiPath program execution.

Finally, you learned how a typical RPA organization would be set up in a UiPath program. This is just the tip of the iceberg, and there is plenty of information about setup and configurations in UiPath documents; please deep-dive and learn as and when the need arises.

Now that you have a basic understanding of UiPath Platform, it's time to get moving to the next topic. Let's learn about the UiPath support strategy, framework, and models in the next chapter.

2
Defining UiPath Support Strategy, Framework, and Models

According to studies conducted by **Ernst & Young (EY)** and Gartner, in recent years, many RPA programs have failed to meet their objectives. One of the primary reasons highlighted is the lack of maturity in RPA maintenance and support operations.

Supporting the UiPath RPA processes in production is the most challenging and most rewarding work, as it helps the RPA **Center of Excellence (CoE)** meet its objectives. UiPath automation program leaders understand that building the bots is easy, but maintaining them is the hard part. Many UiPath programs do not meet the desired ROI because the bots were not supported and managed correctly after production going live; therefore, it is vital that UiPath automation leaders understand the importance of bot support and maintenance as early as the pilot phase.

In this chapter, we will cover the core UiPath Support operation basics. We will begin with an introduction to the concept of UiPath Support, then explain the core constructs of how to run a successful UiPath Support organization, and finally, end with how a support organization delivers value to the overall UiPath RPA program by collaborating with various stakeholder teams.

In this chapter, let's expand the *ABC Insurance Corporation*, an insurance service provider, and a UiPath customer reference. They have set up an RPA CoE and have mature Support and monitoring operations. Let's assume that a new UiPath support team member, Jennifer, has been onboarded to Candace's UiPath team. Let's walk through this chapter's concepts to complete Jennifer's initial part of the UiPath Support onboarding training.

> Note
> The most common mistake that many UiPath RPA programs make is to ignore the importance of support and monitoring functions and just focus on the development of the bots.

In this chapter, we will cover the following topics:

- Getting an overview of the UiPath Support setup

- Understanding the RPA Support strategy, framework, and model

- Learning how to define an efficient UiPath Support policy

- Knowing how to onboard and offboard UiPath bots into production support operations

- Learning about different teams involved during RPA support

- Understanding why and when to reach out to internal and external support teams such as UiPath Support

Let's begin Jennifer's UiPath Support training by getting started with the importance of a support strategy and defining a framework for delivering strategic outcomes.

UiPath Support strategy and framework

The UiPath Support and Monitoring functions are responsible for maintaining and continuously improving the UiPath bots in production, the UiPath platform, and the infrastructure components. In many organizations, deployment (or release management) is also handled by this function.

> **Note**
> The UiPath teams mentioned in this book refer to the internal bot support team that supports the UiPath RPA program at the ABC Insurance Corporation and not the support team from UiPath "the company."

Strategy is the foundation of any successful support operation. Like any other operation, the UiPath Support operation should be built on a sound strategy. It will ensure UiPath Support's operational success and enable the scalability of the RPA operation when needed. The strategy should be flexible enough to adapt to change and respond to organizational demands, as the maturity of the UiPath Support operation increases over time.

Strategies are defined by the goals and objectives to be achieved by the team. It should be aligned with the overall business model.

> **Note**
> In most organizations, when the UiPath pilot program starts, the support and monitoring activities will just be managed by the developers who built them originally. Even in this timeline, it is recommended that you have a UiPath Support strategy to steer the program to success.

The UiPath Support strategy should be defined to align with the **UiPath RPA CoE** goals and objectives and aligned to the overall IT Automation strategy of the company.

The *ABC Insurance Corporation's* UiPath RPA CoE has the following initial set of goals and objectives for its UiPath Support team by the leadership council:

- **O1**: Improve customer experience by having quality support and monitoring in place.

- **O2**: Enable the UiPath Support operation to scale up.

- **O3**: Minimize the downtime of bots and increase the availability of operations.

- **O4**: Optimize resource utilization and reduce costs through continuous improvements.

> **Note**
>
> The UiPath Support objectives defined here are just a sample extracted from the ABC Insurance Corporation to explain the concept, and they may be applicable to all organizations.

The UiPath Support strategy consists of different steps or measures that need to be taken to meet the defined objectives. Let's get into those details in the following section.

UiPath Support strategy overview

A typical support strategy will be defined to meet the objectives. Still, as we do not have infinite resources, we must balance three different dimensions, Time, Cost, and Quality, and get the best possible outcome:

Figure 2.1 – The three dimensions of the support strategy

If we extend the example of the ABC Insurance Corporation, then the RPA CoE leadership team has defined a strategy based on these three dimensions:

- **Service-Level Agreement (SLA)**
- **Total Cost of Ownership (TCO)**
- **Risk**

Additionally, they also defined many tactical initiatives to achieve their goals. Let's see them in detail in the following sections.

> **Note**
> The RPA Support strategy should be defined by a *top-down* approach, where the leadership team defines and sets the direction, but the actual execution happens from *Grounds-up*. Therefore, it is recommended that you have an experienced operational leader as part of the core strategic leadership group.

SLA – time dimension

SLAs are defined at different levels in the UiPath Support program. A process-level SLA will deal with metrics such as volume, average handling time, downtime, and more. Platform- or infrastructure-level SLA will measure availability and capacity. Additionally, they will be applicable to measure the UiPath Support team with metrics such as response time, incident completion, escalations count, and more.

Some of the tactical initiatives used by the ABC Insurance Corporation to meet the objectives related to SLA are listed as follows:

- **T1.1**: Use a technically sound solutions-based **IT Service Management** (**ITSM**) suite such as ServiceNow (used to streamline the enterprise IT support processes, work allocations, SLA calculations, and reporting) in UiPath Support services to help manage UiPath Support operational health and improve performance.

- **T1.2**: Touchless request handling is the most efficient solution in support operations. Implement UiPath Support self-service wherever possible; the service catalog was in place for users to perform credentials unlocks, add a list of emails to UiPath job reports, or rerun failed UiPath jobs.

- **T1.3**: Integrate the UiPath platform with an advanced and mature monitoring and alerting solution that provides 24*7 monitoring for running UiPath RPA jobs with minimum human involvement.

- **T1.4**: Have a centralized bot information catalog solution to help reduce bot downtime by supporting intelligent impact analysis.

> **Note**
> The leading causes of bot downtime or breaks are changes to the business application's flow or **User Interface** (**UI**), inefficient exception handling, and the misinterpretation or wrong understanding of requirements.

TCO – cost dimension

Cost is the most critical metric in many organizations when measuring the success of a UiPath Support program. Just like SLA, cost also has many dimensions of measurement. Overall, the most measured metric is the TCO, which can be scoped to the support organization, the UiPath Support resources, the platform operation run cost, and the monitoring solution cost.

Some of the tactics used by the ABC Insurance Corporation to meet the objectives related to TCO are listed here:

- **T2.1**: Minimize the TCO by mixing an intelligent operating model with internal and external resources executing support operations.

- **T2.2**: Use UiPath unattended robots to run health checks on the various bots and systems they were automating, check the service queue, and send alerts for high-priority issues.

- **T2.3**: Use attended robots that will work along with UiPath Support personnel to reduce the cost of UiPath Support request execution.

- **T2.4**: Use intelligent monitoring solutions to maximize resource utilization, for instance, bots, platforms, external systems, or infrastructure.

- **T2.5**: Generate regular reports on the cost-benefit analysis brought in by running a lean and efficient UiPath Support organization.

Risk level and program management – quality dimension

The quality aspect of UiPath Support revolves around reducing the risks and problems in supporting UiPath applications and platforms. It is one of the least-focused metrics but very critical to the success of the UiPath program. Like SLA and TCO, risk and problem management also have many measurement dimensions: at the bot execution level, the monitoring level, the support resources level, and the platform level.

Some of the tactics used by the ABC Insurance Corporation to meet the objectives related to quality dimensions are listed here:

- **T3.1**: This document risk levels as part of the risk governance, supports handover documentation, and sets up alerts in an automated monitoring system. When a business-critical bot fails in production, a risk alert is sent out to the UiPath Support team, leading to faster issue resolution.

- **T3.2**: Improve the quality of service provided to the end customer of the RPA bots. Implement a system for helping with an automated reply for requests, progress updates on support incident progress, tracking customer satisfaction ratings, and more.

- **T3.3**: Implement a sound knowledge management solution to have all the documentation and references available for the UiPath Support personnel that drastically improves the quality of service provided.

- **T3.4**: Implement associations between incident management, problem management, and change management in the ITSM to improve support quality by reducing repeated issues, which will also improve the UiPath bot and platform availability metrics.

> **Note**
> The UiPath Support strategy and frameworks are defined by the RPA CoE strategy definition committee and are included in the RPA governance initiative. UiPath IT leaders, business sponsors, Enterprise Architects, Support leads, and more will be part of the strategy definition leadership team. The UiPath Support objectives will differ based on organization goals and UiPath Support strategies will be defined in accordance with these objectives.

Most organizations do not even have a dedicated support team when starting the RPA initiative. Therefore, the UiPath Support strategy is not on their radar until the bots in production start failing. Then, they hire or convert existing resources to help with support and monitoring.

It is imperative to understand that this UiPath Support strategy is the foundation of a successful RPA program. I hope the three dimensions discussed in this section should ring some bells when defining a support strategy for your organization.

Now that we have looked at the strategic angle, let us get into details on how to get the strategy results. The UiPath Support framework is the vessel that can link the strategy to the execution. In the next section, let us look at the different components of the UiPath Support framework.

UiPath Support framework

The framework is a set of building blocks that enable the strategy's execution. We will explain this concept in the context of UiPath Support with the help of the ABC Insurance Corporation. Let's look at the UiPath Support framework used by the ABC Insurance Corporation in detail:

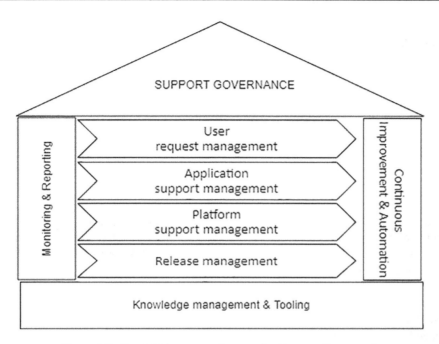

Figure 2.2 – The ABC Insurance Corporation Support framework

- **Governance**: The UiPath Support governance is the most critical component of the overall support framework. The scope of the support governance also covers the monitoring and deployment aspects (in some cases). Usually, it is recorded in a document covering the policies, rules, processes, and reporting metrics used to govern the UiPath Support and monitoring setup. Roles and responsibilities, along with SLAs and escalation paths, are also defined in this document.

  Once the governance document is ready, it must be reviewed by the UiPath RPA CoE leadership team and signed off. The most challenging part of governance is enforcing the governance rules. Hence, usually, the RPA CoE Support manager should take accountability for enforcing the rules in the UiPath Support execution. Additional checks need to be in place to review the enforcement periodically.

- **User request management**: One of the primary functions of the UiPath Support team is to handle the requests (including questions) raised by end users of the UiPath bot deliverables. The framework must define SLAs for handling such requests-based business impact and priority ratings and the channel of communication to the users.

  In the ABC Insurance Corporation RPA support team, let's consider two open requests:

  - The first request is for adding 100 new users for access to the reporting dashboard.

  - The second request is to restart a business-critical job that has financial impacts.

A user request management process should be set up to prioritize the second request before considering the first one. These finer details need to be communicated within the documents related to user request management.

- **Application support management**: This bucket covers the support provided for production UiPath jobs. The process support documentation must define the policy and SLA definition regarding the live bot's support. This section also covers the steps that the Support professional must take once a production bot issue has been notified by the business user or the monitoring team.

> **Tip**
> When strategizing about application support, consider how to monitor the process in the production. The metrics we will be collecting and reporting on should be defined in the PDD. It's common to get to the "release to production" stage and realize that you need to add a great deal of data collection or monitoring hooks to the process. You will want to push these sorts of activities as early as possible in the life cycle.

For instance, consider an ABC Insurance Corporation RPA operation where a production UiPath bot job failed to execute. A failure audit email will be duly sent. Then, the monitoring team will raise the red flag and log an issue with the support team. The level 1 support will refer to the documentation, and once the root cause has been identified, they will escalate this issue to the level 2 support team. This team will handle the web selector fix, and the change management will also be triggered. These kinds of details will all be covered in the governance or operational documents.

- **Platform support management**: The UiPath platform or infrastructure support needs and policies are covered in this section. The UiPath Support team needs to know how to react when the root cause of the issues is related to the following:

a) UiPath platforms such as Orchestrator, Robot, and more

b) Infrastructure such as VMs, databases, elastic search servers, and more

For instance, in the ABC Insurance Corporation, a UiPath production job did not create execution logs in UiPath Orchestrator; the support team checked the procedures in place to categorize the issue as platform-related and then created a ticket with the UiPath product support team to get their assistance in fixing the issue. Such kinds of platform-related support request-handling procedures will be covered in the support documents.

- **Release management**: This section covers the process, roles, and integrations necessary for the release and deployment management of UiPath packages to different environments such as test and production. Integration to change management and deployment channels will be covered, too.

In the case of the ABC Insurance Corporation, the RPA CoE release management policy dictates that the UiPath Orchestrator administrator can only do the releases through an automated DevOps tool (Jenkins) pipeline for approved changes. The releases are scheduled to be deployed during an official release window.

- **Monitoring and Reporting**: This section covers all the aspects of how the monitoring team should be set up. Also, it covers the types and frequency of reports expected from a UiPath Support organization. The governance of all the monitoring solutions deployed, including UiPath Insights and third-party applications such as Kibana, are also covered in this section. In many companies, enterprise monitoring solutions such as Splunk, Datadog, and Prometheus are also used for monitoring and alerting, and they are also scoped into this bucket.

  Reports are critical for a successful UiPath Support operation. They are not just used for reporting on bot performance but also for the impact of the support team on the overall UiPath programs. Support ticket reports, SLA adherence, bot versus process metrics, license usage, and change deployment summary are standard reports generated in the ABC Insurance Corporation UiPath RPA program.

- **Continuous Improvement and Automation**: Continuous improvement can be achieved with support and monitoring process improvements or by building new support and monitoring tools or utilities. In the ABC Insurance Corporation, daily health report generators or license usage alerts are put into place to reduce the support handling time and improve quality.

  The idea is that continuous improvement programs must be incorporated into support team operations to improve how the team operates. This is the best way to keep the support team motivated.

- **Knowledge Management**: The foundation of the UiPath Support framework is knowledge management processes and policies. All of the support-related documentation should be stored in a central repository (such as a Confluence or SharePoint site), and they should be readily accessible to the support and monitoring team members.

  Troubleshooting guides, bot support handover documents, training, and more are some of the typical documentation available in the ABC Insurance Corporation RPA knowledge management repository.

- **Tooling**: To get the best results, it is highly recommended that you automate the support framework with a tool or web portal as much as possible. The framework cannot add value to the program without a delivery channel.

  ITSM applications (such as ServiceNow and Zoho, and more), application life cycle management products (such as Jira and Rally), and monitoring solutions (such as Splunk and AppDynamics) can all be integrated with a dashboarding tool such as Kibana or Power BI to paint a beautiful picture of the UiPath Support team's operational details and platform performance if they are implemented in the right way.

> **Tip**
>
> The framework defined in this section is just a recommendation. The strategy definitions impact the individual component that will be used in this framework, and it takes time to get all the components in place. It is recommended that you have the proper documentation for the components in the Support framework, which can be found in the *Governance* section of the *Knowledge Management* solution.

Now that Jennifer has understood some aspects of the UiPath Support strategy and framework, in the next section, let's look at the details of the support model.

UiPath Support model

The UiPath Support model determines how value is delivered to the end customer. For our purpose, we will focus on how the UiPath Support team addresses support requests. In this section, we will look at the two main topics of a UiPath Support model:

1. **Sourcing model**
2. **Operating model**

UiPath Support sourcing model

Usually, IT procurement and vendor management teams drive support sourcing decisions. A typical sourcing model involves the following:

- **Internal resources**: All the support and monitoring resources are internal to the company that operates the RPA CoE. This model is popular with a UiPath Support function with low maturity, for instance, programs that are just starting with a few bots or have a strict security and compliance policy where external resources are not allowed, such as in a few US federal programs.

- **Combination of internal and external resources**: This is the most popular resourcing model employed in many organizational UiPath RPA support and monitoring programs. Here, the right mix of internal and external contractors is part of the team. One of the significant advantages of this model is its ability to allow the RPA support program to scale up and down as per the changing needs of business and IT. The ABC Insurance Corporation also uses this resourcing model to run its successful UiPath Support and monitoring program.

> **Tip**
>
> Internal, external, and managed services can be onshore, offshore, or nearshore. In the new post-COVID era, resource location is no longer a problem.

- **Managed services**: This resourcing model deals with the complete outsourcing of the UiPath Support and monitoring function to an external service provider. It is gaining momentum for UiPath Support programs that have reached an elevated level of maturity where most support and monitoring functions are streamlined, documented, or even automated. If the risk to the program is low, then this is the most cost-efficient resourcing model for a sizeable UiPath Support operation. Most of the managed services will be operated from a cost-efficient offshore location.

> **Tip**
>
> In recent years, many RPA service providers have been delivering RPA-managed services in an on-demand model, for instance, you pay for the services you use during a particular period as per the need. For instance, managed service can provide level 1 and level 2 support for 5 days a week, help with upgrades, provide monitoring services, and more. It is an evolving space that UiPath Support and administrator students need to be looking out for.

The UiPath RPA program's typical support team structure

Building a support team from the ground up is a very challenging task. Having a good mix of UiPath technical and monitoring experts is vital for a successful UiPath Support organization.

The UiPath RPA CoE team structure is followed by the ABC Insurance Corporation. As this is a highly mature RPA CoE, various roles are involved in its UiPath Support and monitoring operation:

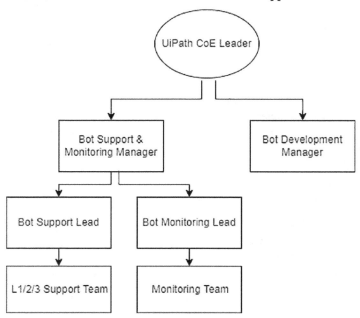

Figure 2.3 – The ABC Insurance Corporation support organization

> **Tip**
> The resources of the ABC Insurance Corporation support team will be used in release management, and they will also be used in supporting and monitoring utility development.

- **Bot Support team**: The core team supports any UiPath Support requests and incidents raised via an ITSM system. A mix of support professionals primarily handles the level 1 and level 2 support tickets. Also, the support team helps with UiPath administration requests, such as user access and bot administration, and maintenance requests such as upgrades and VM cleanups.

- **Bot Support lead**: The support lead, as the name suggests, will be the lead for all of the UiPath Support professionals by providing technical expertise on complex requests, handling communications, handling triages of requests, prioritizing requests, helping with escalations, coordinating with external teams, and helping with reporting aspects of the Support program.

- **Bot Monitoring team**: The UiPath monitoring team will be responsible for monitoring the technical health of UiPath jobs, platforms, and even the infrastructure. This team activity becomes very critical once the volume of the job scales up. Most of the team started with Orchestrator to monitor the jobs and later relied on Audit email monitoring. A more mature team such as ABC Insurance Corporation has integrated UiPath events with enterprise monitoring solutions, and they try to provide 360-degree coverage on all aspects of the UiPath program.

- **Bot Monitoring lead**: The monitoring lead is responsible for leading the monitoring team to mature the monitoring and reporting practice. One of the primary roles of the monitoring lead is to look at coordinating and improving the monitoring coverage of the bots, platform, and infrastructure. It would also involve tracking the metrics needed for the RPA CoE leadership, for example, ROI, bot utilization, job volume trends, and more. In many organizations, UiPath Insights (the analytics offering from UiPath) is, usually, managed by the monitoring leads.

- **Bot Deployment manager**: UiPath deployments are performed on the different environments that were introduced in *Chapter 1*, such as *Test, UAT, and Production*. This role is the cornerstone for ensuring the deployments are performed on time and that all the configuration items, such as assets, triggers, and accesses, are in place for a bot to work once the packages have been deployed. Additionally, this role is responsible for automating the manual deployment process into a more automated deployment process using the **Continuous Integration (CI)** and **Continuous Delivery (CD)** pipeline (which will be covered in a later chapter).

- **Bot Support Automation team**: This is an extended team. Primarily, the team is used to build UiPath Support and monitoring utilities. This can be full-blown automation with UiPath, such as addressing the rerun of job requests, or it can be a simple script that will clean up files in a shared drive tool. For a mature RPA Support and monitoring team, these continuous improvement initiatives are essential for adding value to the program in the long run.

> **Note**
>
> Depending on the maturity of the RPA CoE, these roles could be performed individually, or single resources will play multiple roles. As the organization scales, individuals will specialize in fewer and fewer roles, until ultimately, you'll potentially have several individuals playing several of these roles, and the others will be dedicated individuals. At 5 bots, you'll have one individual playing the monitoring lead role, using about 10% of their time, and you'll have one dev doing the monitoring team role, using about 20% of their time. Once you're at 5,000 bots, you'll likely have a monitoring lead for each business unit and a whole team of Monitoring team folks.
>
> To put this into perspective, in the ABC Insurance Corporation, 200 RPA processes are running on 250 bots; so, many individuals might be assigned to each possible role. On the other hand, if there are only 10 bots in production, then a 2-member part-time support and monitoring team, each of whom plays multiple support roles, should be sufficient.

The UiPath Support operational model

As the name suggests, support operational processes define the management, reporting, scheduling, and referral handling processes for **Business-as-Usual (BAU)** applications. The usual IT application support operating model will also support UiPath Support requests. The following section explains three layers of support in the context of UiPath.

Trifactor Support setup

Here are the three layers of support:

- **Level 1 support**: The UiPath Support team can handle any requests and incidents by following the **Standard Operating Procedure (SOP)** or training documentation available from a knowledge management application. In some organizations, this function is integrated with the IT helpdesk or enterprise support team.

- **Level 2 support**: Usually, when an incident needs an update to the script, the tickets get escalated to level 2 support and internal UiPath developers. They help resolve the incident with a script update, and the ticket gets addressed.

- **Level 3 support**: If the ticket's root cause analysis points to a complex solution, we need external IT or business application teams (including the UiPath Support team) to solve a bot production issue. Then, they are addressed by this on-demand level 3 support team, which is usually coordinated by the support lead.

> **Note**
>
> In the case of the ABC Insurance Corporation, they also had level 0 UiPath Support. Level 0 covers the RPA support service catalog that lists all the self-services available to end customers such as password reset or rerunning a failed job. As the support model matures and moves toward touchless mode, self-service catalogs are getting popular with time.

The following diagram depicts how a support process works on a conceptual level.

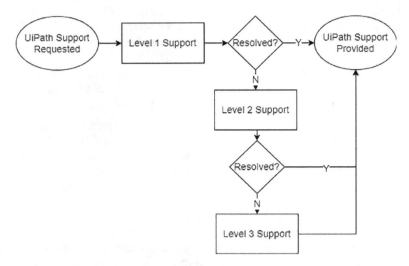

Figure 2.4 – The operational support process

At a high level, the process described in the preceding diagram is applicable to all highly mature UiPath Support organizations. First, the level 1 support team will triage a support request that gets placed, and the request will be resolved based on the agreed SLA. If the request cannot be solved with canned troubleshooting steps or a training guide, then it gets escalated to the level 2 support team, where UiPath developers will become involved to get the issue resolved.

> **Note**
>
> The official UiPath support team will only get involved if there is a support contract between the customer and UiPath. There are professional UiPath Support service packages starting from "Standard" to "Premium," "Premium Plus," and "Platinum." Customers need to opt for this option to leverage the UiPath Support team's expertise. Please refer to this link for more information: `https://www.uipath.com/support/packages-options`.

If the issue was not solved by level 2, then it gets escalated to the level 3 support team, where resources from the external team and even UiPath Support experts get involved to resolve the issue.

> **Note**
>
> In a particular scenario at the ABC Insurance Corporation, the RPA support team could not solve a support issue even when the UiPath Support team got involved. The issue was logged as a product bug. Later, a software patch was developed by the UiPath product team. Then, the patch with the issue fix was applied to Orchestrator with an upgrade.

Maturity model

UiPath Support's operational maturity model can be defined based on operational excellence, knowledge management capability, automation score, monitoring setup maturity, and more. Usually, it is defined to help the UiPath Support program measure its current maturity and plan to improve its operations. The following model was deployed in the ABC Insurance Corporation's UiPath Support team:

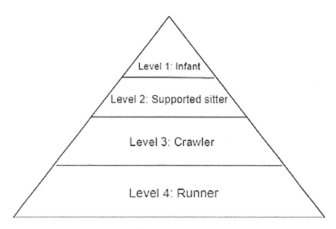

Figure 2.5 – Support operational maturity model

Here are the maturity levels in more detail:

- **Maturity Level 1**:

  - **Roles and responsibilities**: `Not defined`; responsibilities are picked up on an on-demand basis

  - **Service delivery process**: `Not defined`; an adhoc process setup makes it difficult to meet SLA targets

  - **Documentation**: `No documentation`; the support teams rely on developers' knowledge

  - **Automation**: `None`; all the support and monitoring activities are performed manually

  - **Monitoring**: `Basic`; UiPath robot jobs are monitored by audit emails and orchestrator logs

- **Maturity level 2**:

  - **Roles and responsibilities**: `Defined`; all the support and monitoring team members are educated on their roles and responsibilities.

  - **Service delivery process**: `Defined`; the processes are set up but not integrated into the enterprise ITSM application.

- **Documentation**: `Available`; the support teams use some of the support troubleshooting guides, and a few bot process training documents are available.
- **Automation**: `Semi-automated`; there is some automation available for support processes. Custom utility or bot solutions are built to help with support and monitoring operations.
- **Monitoring**: `Bit advanced`; monitoring solutions, such as the UiPath Insights or Kibana dashboard, are in place. The team is trained to use them for monitoring and reporting purposes for all bot jobs.

- **Maturity Level 3**:

  - **Roles and responsibilities**: `Refined`; all the support and monitoring team members are cross-trained and have equal knowledge to support and monitor the UiPath platform and processes.
  - **Service delivery process**: `Automated`; the processes are integrated into the enterprise ITSM application, and all the RPA support services delivering processes such as a) intakes, b) processing, and c) reporting are automated. Additionally, self-service for UiPath Support is available for end users.
  - **Documentation**: `Version Controlled`; the support teams have access to all the support and monitoring documentation, which is fully documented. They are frequently updated and versioned, too.
  - **Automation**: `Mostly automated`; many automations are available for support and monitoring. Very few tasks need to be performed manually by the support and monitoring team.
  - **Monitoring**: `Advanced`; multiple monitoring solutions to monitor the UiPath process, platform, and infrastructure are integrated and available for the support and monitoring team. Business users are also trained to build and use reporting dashboards.

- **Maturity Level 4**:

  - **Roles and responsibility**: `Refined`; all the support and monitoring bots and team members know their separation of duties and work in harmony.
  - **Service delivery process**: `Fully Automated`; all the user requests are serviced through the self-service portal or application, and complete visibility is provided on the submitted requests.
  - **Documentation**: `Complete`; interactive documents and training videos are available, and the support teams have access to all the support and monitoring videos and documentation. They are integrated into the support interface and can be accessed instantaneously.
  - **Automation**: `Completely Automated`; level 1 support and monitoring requests are automated by bots and other solutions.

- **Monitoring**: `Very Advanced`; self-healing solutions are advanced enough to trigger jobs that will fix production issues without any human intervention. Customized reports and dashboards based on personas are built automatically on demand.

> **Note**
>
> The ABC Insurance Corporation is currently in level 2 on this scale. Many UiPath RPA support programs are currently in the level 1 or level 2 maturity levels. This is a clear indication of the demand for quality UiPath Support and administration professionals in the job market. Level 4 is the ideal state, and I believe that advances in machine learning and other artificial intelligence fields will make this possible in the near future.

I hope this was interesting learning for Jennifer. In the next section, let us look at the details of the support policy.

Support policy

UiPath Support policy should be documented, approved, and shared by the RPA CoE leadership team. Policies are rules that must be enforced to provide UiPath Support in the most transparent way possible without any bias and help with standardization and audits.

The support policy document defines the procedures, roles, and responsibilities required to support UiPath processes running in the production environment. It covers various aspects tending toward the roles, responsibilities, processes, and tools that are needed for the following activities:

Figure 2.6 – Support policy building blocks

- **Support Requestion**: UiPath Support requestion is an important step. The policy document should set the direction for different stakeholders such as business users, external teams, the RPA leadership, and more to submit requests to the UiPath Support team. Here are some essential aspects of this step. Let's consider the ABC Insurance Corporation's UiPath Support intake policies to explain this better:

  - **Channel**: Different channels such as email, web portal tickets, phone calls, and chat were provided to UiPath RPA end customers to log support requests.

- **Catalog**: A portal with a UiPath Support service catalog that has few self-service options and regular UiPath Support request services was implemented.

- **Qualification (scoping)**: Triage policies were set up to check the validity of the request and to decide whether the requests and issues can be handled by the internal UiPath Support team or an external team that needs to be involved, too. Once qualified, an acknowledgment needs to be shared with the requester.

- **Categorization**: Clear categories to bucket support requests versus monitoring requests and incidents into UiPath bot process issues, platform issues, and infrastructure issues were introduced so that the right team can handle the issues or requests quickly.

- **Prioritization**: As limited support resources were involved in UiPath Support, clear prioritization policies were put in place to bucket tickets into emergency, high, medium, and low categories based on business impact analysis.

> **Note**
>
> Usually, support requests start with ad hoc requests, and it is vital to streamline the intake process. Otherwise, as the operation matures, it could be unmanageable and lead to a bad customer experience and missed SLAs.

- **Support Servicing**: Policies around the core UiPath Support services are this policy document's crux. These policies will ensure that all the logged requests and incidents are addressed by the UiPath Support team competently and professionally. Let's consider the ABC Insurance Corporation's UiPath Support servicing policies to explain this better:

  - **Request**: All qualified requests need to be handled by the level 1 support team and need to be closed within the agreed SLAs.

> **Note**
>
> Different bot processes might have different SLAs associated with them. For example, a bot that handled sending checks out might need to be fixed by a 4:00 P.M. mail deadline, whereas a bot that sends out a weekly report could be fixed in 4 days. The SLAs need to be captured and documented in the PDD stage and propagated into operations.

  - **Incident**: The incident needs to be bucketed into level 1, 2, or 3 types of incidents, and it needs to be handled as per the agreed SLA. Additionally, proper communication needs to be provided back to the requester regularly.

  - **Problem**: Automated problem identification from frequent incidents was implemented, and integration into the change management process was implemented.

- **SLA**: SLAs for different aspects of support and monitoring were reviewed every quarter (or on an on-demand basis), and agreed SLAs were communicated to the stakeholders.

- **Escalation**: Clear guidance for any escalation path was published, and handling procedures and SLAs were updated based on the business impact.

- **Knowledge management**: Knowledge resources such as documents, video recordings on best practices, bot process guides, and everyday request items were frequently updated on the fly and used during the support servicing step.

> **Note**
> UiPath Support serving is the core responsibility of the UiPath Support team, and complete visibility and notifications need to be provided to the requester on the current state of the request. Usually, it is recommended that you integrate with an ITSM solution for better outcomes.

- **Support Reporting**: The reporting of support activities is also a critical step, and there are a few aspects that will be discussed here. Let's consider the ABC Insurance Corporation's UiPath Support reporting policies to explain this better:

  - **Reporting types and schedules**: Reports were grouped based on user groups such as the business team, the UiPath RPA CoE leadership team, the UiPath Support team, and more. Some of the reports, such as the platform health report, were shared every week, whereas bot performance metrics and ROI reports were shared every month.

  - **SLA breaches**: SLA breaches and business impacts were captured in reports so that improvement measures can be placed to improve UiPath Support and monitor operations continuously.

  - **Customer Experience**: UiPath customer feedback ratings were collected at the end of every support ticket resolution. This is an important metric to track to grow the UiPath practice in the organization.

> **Note**
> UiPath Support operation performance needs to be tracked with the help of a reporting application put into place by the leadership team, to fine-tune the support strategy and resourcing needs, and plan continuous improvement initiatives.

Next, let's discuss support alignment details that needs to be recorded in the UiPath Support policy document.

Support alignment

At a high level, the support policy needs to define how the support team must align with different enterprise functions, such as the following:

- The business or product team
- The RPA development team
- The business application helpdesk and its support
- Change management
- Deployment and release management
- Monitoring
- Communications
- The infrastructure team
- The data management team
- Risk management
- IT security, audit, and compliance

> **Note**
> UiPath Support policies should be defined with proper alignment to existing governing policies in other teams who request or provide support to the UiPath PRA programs. It is good to share the policy document with these partnering team leadership teams in order to align SLA and reporting requests.

Next, let's expand Jennifer's understanding of the UiPath Support life cycle.

Support life cycle

There are two core terminologies where many support students or beginners get confused:

- **Bot**: This refers to a virtual worker who will operate in a virtual environment and occupy resources, including a license, in the UiPath platform. The bot can execute multiple processes (in the foreground or the background).

  A typical bot life cycle, such as a human employee, begins with an onboarding process where the bot is configured into a machine or machine template, and access to resources such as applications, shared drives, mailboxes, credentials, and more is assigned. Then, once the bots are ready, they are operationalized to execute one or more processes at various points in time; the bot will be continuously monitored for health and utilization. Generally, they get repurposed to execute different processes. If the demand for the bot decreases due to changing business conditions and other internal or external factors, the bot license is unallocated, and the resources are released.

- **Process**: On the other hand, a process is an automation program deployed on an RPA bot to execute business operations.

The bot process life cycle is critical to understand; let's see the different components of this life cycle in more detail:

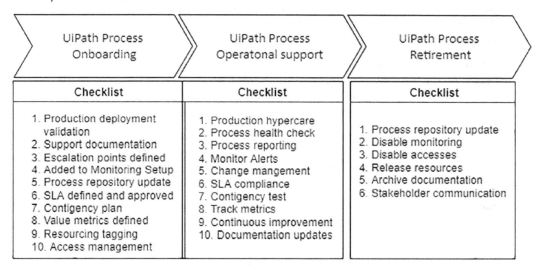

Figure 2.7 – The bot process life cycle

UiPath process onboarding

Process onboarding is the step that happens once the UiPath package is ready to be deployed to production. These are the prerequisite steps that need to be performed to execute the bot process successfully. There are different tasks involved in the process onboarding step of the ABC Insurance Corporation UiPath Support operation:

1. **Production deployment validation**: Production validation and sign-off provided by the business and IT stakeholders is the prerequisite for any bot process onboarding. The validation checklist is maintained, and once all the items have been checked, the validation sign-off is provided by the business stakeholders.

2. **Support documentation**: The UiPath RPA developer of the bot process will prepare the support handover documentation and provide training to the UiPath Support team when the bot process is onboarded into production support.

3. **Escalation points defined**: The support team must be provided with the business and developer contacts for the bot process in case of issues or the need for clarification during production support. It is recommended that you have this information documented in the support handover documentation and shared in a knowledge management solution.

4. **Added to the monitoring setup**: Support and monitoring teams should ensure that the new onboarding bot process is added to the existing monitoring alerts and reporting setup. Custom alerts for this new process can also be set up during this step.

5. **Process repository update**: Usually, the process repository has all the bot process information such as the job names, the trigger information, the application involved, the SLA, and more. It needs to be updated with the new bot process information.

6. **SLA defined and approved**: Business stakeholders need to provide the SLA for the bot process, for example, the bot process should send the audit report before 7 P.M. on the same day or the bot needs to process 95% of the transaction successfully. These SLAs need to be discussed between the business, IT, and RPA stakeholders and signed off.

7. **Contingency plan**: If a bot process stops working due to a planned outage or an application issue, a contingency plan should be put into place to handle the business operations manually or a workaround until the bot process is fixed. These plans should be documented in the process handover document, too.

8. **Value metrics defined:** Bot process metrics are vital for calculating and reporting on the value provided by the bot process. Different metrics that need to be tracked on the bot process, such as the volume, success rate, average execution time, and more, are defined at this stage.

9. **Resourcing tagging**: The physical and virtual resources needed for the bot processes must be defined. Relevant teams need to be involved in requesting those resources for the bot process, for example, AWS S3 bucket access, an Office 365 license, and more.

10. **Access management**: The bot process might need access to various business applications, shared drives, or document management systems. This access needs to be requested and tested before deploying the bot into production.

> **Note**
>
> It is recommended that the onboarding checklist is included in the pre-go-live meeting, and its approval should be the prerequisite for triggering the bot release and deployment process.

UiPath process in operations

Once the UiPath bot process has been deployed to production and the processes are supported in hypercare mode, there are certain best practices that the UiPath Support team needs to follow to confirm and sustain the bot processes' operational health. These are the steps that need to be performed for the successful execution of the bot process. There are different tasks involved in maintaining process operational health in the ABC Insurance Corporation UiPath Support operation. Let's discuss the list:

1. **Production hypercare**: Once the bot process is in production after business validation, the UiPath Support team will move the process into a hypercare phase where responsibility gracefully transitions from development to support, which means that the support personnel gets a feel for how the bot is working and that the development personnel takes it on the chin if the bot is brittle and breaks all the time. The newly deployed bots are monitored extensively compared to other existing bots in production. Hypercare exit criteria should be defined during the bot onboarding phase.

2. **Process health check**: Continuous effort in monitoring and improving the bot process health will be in place. It will include reviews of audit logs, reports, dashboards, and custom health reports of the bot process.

3. **Process reporting**: Proactive monitoring and reporting will be in place to update the successful and failed jobs in production. Custom reports based on personas are built, and accountability is given to the monitoring team.

4. **Monitoring alerts**: Alerts are set up to help the monitoring team proactively find issues in the jobs running in production. Orchestrator email alerts, custom email alerts, alerts generated by enterprise monitoring tools, and more are some of the alerts that the support team can request and configure to enhance monitoring.

5. **Change management**: There must be a tight integration with the change management process to handle bugs or enhancements in the production of existing bot processes. Rules are in place to escalate high-priority incidents and fix them quicker.

6. **SLA compliance**: Process SLA compliance is one of the most important metrics to track and report. This step makes sure that the SLAs defined and agreed upon are being met by the bot process. These reports are helpful to implement process improvement initiatives, too.

7. **Contingency test**: It is best to execute the contingency plan and run this test during outages to see how business-critical bot processes can be handled manually.

8. **Track metrics**: There are different metrics such as the process volume, the average handling time, the success and failure rate, and more that need to be tracked on all production bot processes.

9. **Continuous improvement**: The support team implements continuous improvement initiatives such as automated or self-help on the process, improved monitoring, and more on all bot processes in production to sustain the performance and improve the efficiency of the bot processes.

10. **Documentation updates**: Support documents on production bot processes should be periodically updated with the latest SLAs, metrics, and troubleshooting details that the support team members encountered in production support.

> **Note**
> Maintain and update bot process information such as support documents, change records, and metrics such as SLA updates. Once these bot processes have been deployed in production, it is often an overlooked but critical step for successfully supporting the bot process during its lifetime.

UiPath process offboarding

The UiPath Support team needs to follow a few steps to offboard the bot process from the production environment once the UiPath bot process has run its course and when the (business, IT, and RPA) leadership team decides to replace the bot process with a new one or a better solution.

Here are the different tasks involved when offboarding the bot process in the ABC Insurance Corporation UiPath Support operation:

1. **Process repository update**: Once the process has been offboarded, the UiPath administrator will disable triggers, assets, and processes in Orchestrator. In addition to this, RPA process details stored in the knowledge management document or application will be marked as inactive and then sequentially retired after some time.

2. **Disable monitoring**: Both process- and enterprise-level monitoring for the bot process and its resources need to be disabled once the process has been retired.

3. **Disable accesses**: Business application and cloud infrastructure accesses provided for the bot to execute the process need to be revoked once the bot process is sunset.

4. **Release resources**: Resources including hardware, cloud, and software assets (including licenses) tied to the retiring bot process should be released.

5. **Archive documentation**: The bot process documentation in the knowledge management system needs to be archived for future reference.

6. **Stakeholder communication**: Formal communication with different IT and business stakeholders involved in the bot execution (covering the primary and tertiary contacts) needs to be sent out once the process is marked as inactive.

> **Note**
> When the UiPath Support team releases the resources, such as VMs and licenses, it should also update user access for the bot accounts to the business applications and shared drives. These two items are often missed and caught during IT audits.

The following section will introduce Jennifer to different aspects of the extended UiPath Support setup.

Extended support setup

The core offering of the UiPath Support team is the production support provided to the bots executing business transactions. To enable a successful RPA operation, the UiPath Support team must constantly communicate to request or provide information to both the internal development and business teams or even the external vendor teams.

Having a good collaboration channel of communication is one of the critical indicators for the UiPath Support team to meet its objectives and improve the team's value proposition:

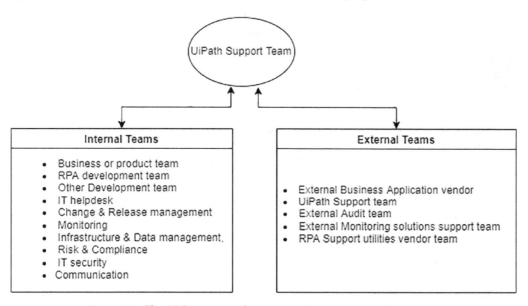

Figure 2.8 – The ABC Insurance Corporation Support team collaboration

Internal support

Internal support refers to the different teams in the same organization; let's say in the case of the ABC Insurance Corporation, the UiPath Support team will reach out to different internal teams at various points in time to request information or assistance to address an open support or monitoring request. They are listed as follows, so let's look at them in detail:

- **Business or product team**: Usually, the support team reaches out to the business or product team to clarify specific requirements or update them on SLA breaches. There are various scenarios where collaboration is needed with the business analyst or product owner to find an alternative solution for an issue or help execute a failed business transaction manually.

- **RPA development team**: A direct communication line with the UiPath Development team is critical to handle level 2 and level 3 kinds of support tickets. Clarification regarding bot process documentation is a common occurrence in production support.

- **Another development team**: If there are other product integrations, such as SAP or Salesforce, to UiPath bots, then a direct communication channel becomes necessary to inform them of any potential downstream or upstream impact due to a UiPath bot issue. They will also be crucial during complex level 3 support ticket resolutions.

- **IT helpdesk**: Usually, the IT Helpdesk or enterprise support desk redirects level 1 support tickets from end users. Having a support requesting process will be beneficial if the support team needs some help to onboard new team members or bots.

- **Change and release management**: Support team members need to be communicated on upcoming changes on production bot processes or any new bots that are planned to be released for support; hence, a communication line with the UiPath change and release management team is crucial for production support.

- **Monitoring**: Mature organizations have enterprise-level monitoring teams that monitor all applications across the IT landscape. It would be good to know the best practices to integrate UiPath monitoring to the enterprise level and get notifications of any failures in the business applications used by bots in production.

- **Infrastructure and Data management**: In many cases, UiPath bots might be affected by changes to the underlying infrastructure or data. It is recommended that you have contacts with the cloud infrastructure or SQL database management team to resolve issues during support or upgrades.

- **Risk and compliance**: In highly regulated industries such as finance and insurance, risk and compliance rules are constantly changing, and it will be good to have contact with the risk team to understand the impacts of those changes on the bots in production. This relationship will also be beneficial to prepare you for internal or external audits.

- **IT security**: The UiPath Support and administration team might have to reach out to the IT security team for approval to set up new UiPath platforms or access bots on business-sensitive applications.

- **Communication**: The UiPath Support and administration team might reach out to the enterprise communication team to share important communication regarding high-priority RPA bot incidents with broader business impacts, upcoming changes to the bot's platform, or downtime notices.

> **Note**
> There might be other internal teams (not mentioned in this section) involved in UiPath Support, and it is different based on how the UiPath RPA program is set up in a particular organization.

Now that we have some idea of how internal support teams and UiPath Support collaborations work, in the next section, we will investigate the impacts of the external support teams.

External support

External support refers to the different teams outside of an organization that host the UiPath RPA CoE. Let's say, in the case of the ABC Insurance Corporation, the UiPath Support team will reach out to different external teams at various points in time to request information or assistance to address an open support or monitoring request. They are listed as follows, so let's look at them in more detail:

- **External business application vendor**: UiPath bots execute automation on business applications built by external vendors such as SAP and Salesforce. They have an extended ecosystem that relies on third parties for solutions such as cloud APIs and plugins used with UiPath.

  UiPath Support needs to have direct communication channels with those teams to address bot or platform issues. Additionally, the UiPath Support team needs to communicate during planned upgrades and patching exercises on those external applications.

- **UiPath Support team**: Most UiPath customers have support contracts with the UiPath product support team. It is vital to understand how to request help from them by raising requests for, usually, level 3 tickets during upgrades or new product installations. Having access to the UiPath Support team becomes critical for large-scale mature UiPath Support operations.

- **External audit team**: The UiPath Support team might work with external IT audit teams by providing reports and access to the UiPath platform and processes. Understanding the requests from these audit teams will help prepare for those audit operations.

- **External monitoring solutions support team**: Many UiPath customers are integrated with external monitoring solutions such as Splunk, Microsoft Power BI, and Kibana. The UiPath Support team should log issues with those external teams as needed.

- **RPA support utility vendor team**: If there are third-party tools used in building custom utilities or portals used for UiPath Support operations, it is good to know how to request support.

Note

There might be other external teams (not mentioned in this section) involved in UiPath Support, and it differs based on how the UiPath RPA program is set up in a particular organization. It is a good list for students to know to prepare well for UiPath Support and monitoring roles.

In the last section of this chapter, let's see how these concepts were used to mature the ABC Insurance Corporation's UiPath program.

ABC Insurance Corporation UiPath Support strategy, framework, and models in action

The ABC Insurance Corporation's UiPath pilot program was started in 2019 with just a two-member developer team, and within 6 months, they were saving 1 million dollars in operational costs for the company with 20 bots and 15 processes in production. Within a short span, the RPA team grew to 10 members. The team comprised 8 UiPath developers, 1 development lead, and 1 UiPath CoE manager.

Due to the high **Return of Investment** (**ROI**), a dedicated bot support engineer was hired for the team to support and monitor the 20 bots in production. To build on the value proposition of this UiPath automation program, an automation strategic committee was set up, and several objectives were defined for the UiPath RPA CoE. One of the key objectives was to enable the UiPath Support operation to scale up the automation program. This objective brought out the importance of the support and monitoring teams' contribution to add more value and sustain the results.

Then, they started to formulate various strategic and tactical initiatives that were aimed at reducing the time, the cost of ownership, and the risk levels. Then came the road map that formulated the suitable framework, policies, and sourcing model for the entire UiPath RPA program, and the subset covered the support organization, too.

These initiatives helped the UiPath CoE to scale up faster in the next 6 months, By December 2019, there were 70 bots in production, saving close to 5 million dollars in operational costs, with the same-sized development team and an additional three-member support team who played different roles in the UiPath support and monitoring team. The leadership team met at the start of 2020 to tighten the policies that covered different topics, including how to onboard and offboard bot processes, bots, and resources to the bot support team. Smart resourcing options were adopted that enabled the CoE team to scale its operations in a short time.

Additionally, a maturity framework was defined and approved to measure the current state and plan the future state of the UiPath CoE, which included the bot support team operation, too. There were many escalations on the bot support team by not meeting the SLA when external support and business stakeholder teams were involved. So, a focus group was set up to close this gap and ensure better collaboration with the team.

With these changes in place, the ABC Insurance Corporation UiPath CoE grew to 300 bots that were operating 150 bot processes, saving more than 20 million dollars in cost by the end of July 2020, with a five-member support and monitoring team.

Jennifer was hired in January 2021, and the initial training in the support strategy and policies was covered in different sections of this chapter.

Now that we have walked through all the sections of this chapter, it's time to recap our learnings in the following section.

Summary

The UiPath RPA support and monitoring operation is one of the critical tangents for the success of organizational UiPath RPA programs. A UiPath Support program needs to be designed with good fundamentals such as strategy, framework, policy, and more. That is this chapter's core message if you are if you are aspiring to work in UiPath Support and administration roles.

Jennifer started with an understanding of the importance of having a UiPath Support strategy and framework for a UiPath Support operation. Strategies are defined to provide direction to the UiPath Support team to meet the objectives set up by the RPA CoE leadership team. On the other hand, the framework defines all the components needed to execute the strategy.

The following section covered support resourcing and operating models and the support operational maturity model. Then, we covered different elements of the support policy and its scope during requests, handling, and reporting.

This was followed by explaining the UiPath Support life cycle, which covered both the bot and process life cycles, and finally, we ended with the importance of collaboration with internal and external teams.

Throughout this chapter, we used the ABC Insurance Corporation and Jennifer's training persona to explain all the concepts. Finally, the journey of the ABC Insurance Corporation's RPA program was detailed to give you a practical implication of how different topics detailed in the chapter come together to define and scale the operation of a UiPath support and monitoring team.

I hope this chapter was informative. In the next chapter, we'll discuss more on UiPath Support operation enablers.

3
Setting Up UiPath Support Enablers

The previous chapter covered all the core UiPath support operation concepts. Still, for any successful UiPath support operation, we also need an ecosystem for enablers that adds value to an effective UiPath support organization. Enablers in this context mean the capabilities, functions, and tools that will enhance and improve the UiPath support operations. The previous chapter highlighted all the enablers under the *Support Framework* section.

We will now build on what was introduced, and in this chapter, we will provide a *how to set up* guide for these UiPath RPA support enabler components. You will learn to set up UiPath RPA monitoring, reporting, and deployment functions to enable the core UiPath support offerings. The chapter also covers a knowledge management and continuous improvement setup, covering custom support tools and utilities to enhance UiPath RPA support organizational performance. The book's subsequent chapters cover the details of each enabler component and operational challenge.

It is essential to have open channels of communication and collaborations with all the stakeholders of the enablers. All the enabling teams should work in a synchronized way to deliver value and meet the goals set up for the UiPath support organization. This sync between teams is a significant challenge in many UiPath support organizations. Understanding these concepts will prepare you for the upcoming challenge in UiPath support and administration and enable you to excel in your position.

We covered the ABC Insurance Corporation UiPath program journey in the last chapter; the program couldn't have scaled up and sustained its value without the help of the support enablers. The support enabler components discussed in this chapter are already implemented in the ABC Insurance Corporation UiPath CoE. It is vital that all new support team members know about these components before they can start working on the UiPath support assignments. We will use Jennifer's support training persona to explain the concepts throughout the chapter.

> **Note**
>
> The enterprise release management team usually manages UiPath deployments in many organizations. However, the concepts discussed here are valid even if the deployment team is outside the UiPath RPA CoE.

Here is what you will learn as part of this third chapter:

- Different UiPath support enabler components

- Best practices in setting up, monitoring, and reporting

- How to set up the UiPath deployment process

- How to set up a continuous improvement process for the UiPath support organization

- How to set up a knowledge management solution for enabling the UiPath support organization

Let's get started with an overview of UiPath support enabling components. Then, we will deep-dive into how to set up this UiPath support ecosystem to enable a high-performance UiPath support organization.

UiPath RPA support enabler components

When an organization reaches a certain maturity in its UiPath RPA journey, then the importance of UiPath support enablers is sensed. Most of the enablers are usually operated and located as part of the UiPath program or CoE, but in some cases, they are also used by specialized teams.

Enablers are highlighted in the ABC Insurance Corporation's support framework introduced in the previous chapter. Let's explain them in detail to Jennifer.

> **Note**
>
> The RPA industry is poised to grow faster in this decade, and many organizations will realize the importance of the UiPath support enablers as they develop their automation practice. These enablers will be obligatory capabilities and features that a UiPath support organization would need to operate at short notice.

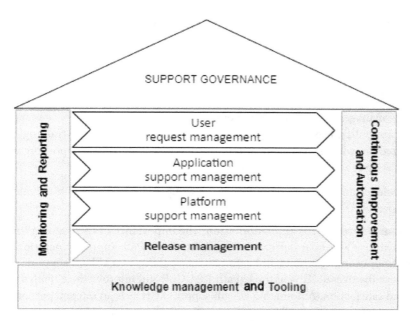

Figure 3.1 – UiPath support framework

The enablers are listed in the order of importance they play in improving the UiPath support operations and adding value to the RPA UiPath support team to march toward the goals and objectives. Let's understand why an organization needs these enablers first before discussing the setup process:

- **Monitoring and reporting**: Monitoring the bot process and resources is the most critical enabler. UiPath audit logs and email notifications are enabled by default in the core UiPath Orchestrator. They form the monitoring center even for a less mature UiPath CoE or program. The value of the UiPath operation needs to be measured to sustain the funding that flows into the program. Hence, it is vital to have a solid monitoring setup in place.

  Reports are a complementary way of monitoring. A measured value needs to be translated into reports customized for specific audiences or personas. Having a good reporting setup for the UiPath operation will ensure RPA program survival and help the operation expand.

Note

Monitoring and reporting are the single most neglected components in UiPath programs. It is tempting to put this off and solve it later. Resist this temptation. Monitoring and reporting are easy to get right if you do them from day one but get progressively harder the longer you put them off.

- **Release management**: Having organized release management will ensure that the UiPath deployments are performed on time and thrive. The release information is communicated to the respective stakeholders before and after deployment.

  Release management is a mandatory enabler present in all UiPath programs, but the maturity of its capability will determine the UiPath support organization's sustainability and growth.

- **Knowledge management**: UiPath support organization cannot operate without a sound knowledge management policy. Having the UiPath support documents and other training materials in a centralized knowledge management repository is essential for bot support organizations to service existing requests and improve the bot support service results.

  If the knowledge management repository is not set up, it will be impossible to scale the UiPath support operation, leading to SLA breaches.

- **Continuous improvement and automation**: The importance of the ongoing improvement process is always an underappreciated capability for UiPath support operations. Having a good continuous improvement and automation capability in a UiPath support organization will improve the overall KPIs of the UiPath RPA CoE and improve the UiPath support team morale and satisfaction. Automating UiPath support activity is an integral part of continuous improvement initiatives that will be a high-potential enabler for any UiPath support organization.

- **Tooling**: Tools and applications to operate the UiPath enablers are vital for the UiPath support operation's success. To streamline the enabler process and their ease of use, having a good set of enabler applications, such as UiPath Insights, ServiceNow, Jira, Rally, and Splunk, are critical capabilities to operate the UiPath support organization efficiently.

> **Tip**
> The enablers listed in this section are the standard components used in different UiPath RPA CoEs and programs. There might also be specific enablers such as auditing or compliance for highly regulated industries.

Now that we have understood the importance of the UiPath support enablers, let's look at the setup details of these enabler components in the following sections and walk Jennifer through the details.

UiPath monitoring setup

The monitoring setup for a UiPath program has multiple levels involved. In the ABC Insurance Corporation, we have three level setups for monitoring:

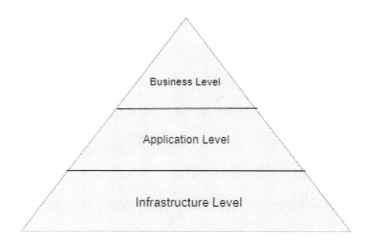

Figure 3.2 – ABC's UiPath monitoring framework

The governance of all the monitoring solutions deployed, including UiPath Insights and third-party applications such as Kibana, are also covered in this section. In many companies, enterprise monitoring solutions such as Splunk, Datadog, and Prometheus, are also used for monitoring and alerting, and these are also scoped into this bucket.

Business level

The business stakeholders of the UiPath RPA program will be interested in monitoring the bot process performance regularly. Provisions and support setup needs to be in place to facilitate their requests.

The ABC Insurance Corporation RPA CoE uses an out-of-the-box monitoring solution from UiPath called **UiPath Insights**. UiPath monitoring can also be configured on custom applications such as Microsoft Power BI, Tableau, and Kibana. Let's look at the monitoring dashboard set up in the following section:

- **Monitoring dashboards**: Business stakeholders of the UiPath RPA process need dashboards to track the bot process performance. The metrics are associated with bot process **Key Performance Indicators** (**KPIs**), SLA tracking, and cost savings. These dashboards are a vital source of information to make informed business decisions. Let's look at the high-level process of ABC Insurance Corporation's monitoring dashboard setup:

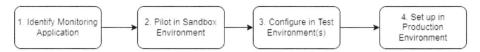

Figure 3.3 – ABC Insurance Corporation's monitoring dashboard setup process

I.   Before we set up these dashboards, we need to identify the primary monitoring application used for monitoring.

II.   Once the decision is made on the monitoring solution, the solution will be piloted in a sandbox environment. A few sample business monitoring dashboards will be picked for the pilot. Once the proof of concept works, the Test and **User Acceptance Testing (UAT)** environments will be set up. Then, the beta test user needs to start using the dashboard, and the solution is finalized or changed.

III. Once all the changes are in place and validated, the UiPath monitoring dashboard can be rolled out to production.

IV.   In the case of ABC Insurance Corporation's UiPath dashboard, there are individual dashboards for all the UiPath processes in production and a consolidated dashboard on function areas such as finance, IT, and HR.

Figure 3.4 – ABC Insurance Corporation's monitoring business dashboard for Processes

---

Tip

Multiple UiPath monitoring dashboard applications are being used in a few organizations based on the business team preference.

- **Business audit email notification**: Business stakeholders of the UiPath RPA process need to be notified of the status of the UiPath jobs in production. UiPath script developers will need to configure this email notification as part of the process automation steps. In many organizations where business stakeholders are very interested in understanding a UiPath job's status, the UiPath CoE support policy dictates that audit emails need to be configured when a bot process is developed. It will be part of the business requirements and documented in the **Process Design Document (PDD)** and production support documents. This kind of policy will force developers to follow a uniform way of scripting the audit email notifications to communicate the success and failure status of a UiPath job in real time.

   There might be scenarios where only handpicked critical processes must be notified to a business. It is usually configurable based on process criticality, using an asset or configurable item.

   The email template and subscriber list need to be configurable items to accommodate future changes in these scenarios.

> **Tip**
> One of the most common requests that UiPath's monitoring team deals with is adding or removing business stakeholders to these audit emails for different processes; hence, the knowledge of audit email configuration is critical for any UiPath monitoring and support professional.

Application level

UiPath support and monitoring teams will need different options to monitor the health of the active UiPath process and jobs in production. To explain the setup, let's discuss various features used in the ABC Insurance Corporation's UiPath COE:

- **RPA monitoring dashboard**: Every UiPath process in production should have at least one representative in your application-level monitoring dashboard. These dashboards can be hosted along with the business monitoring dashboards, and they are configured during the bot process development process. The requirements of the dashboards are part of the business requirements, and the process metrics, such as total volume, average handling time, and success and failure rates, can be tracked for different time intervals such as for a few hours, days, weeks, and even months. The UiPath monitoring team mainly uses it to monitor the bot process run status regularly. When a bot process is promoted to the production environment, its corresponding monitoring dashboard is also available in the monitoring production environment.

- **Application failure alerts**: There are two primary options for alerts:

I.   Application email alerts for faulted jobs from Orchestrator can be used if **Email Setup** is configured correctly under **Tenant | Settings | Mail**. This will be the primary option to check for job faults by the UiPath monitoring team, and the support team can act on them:

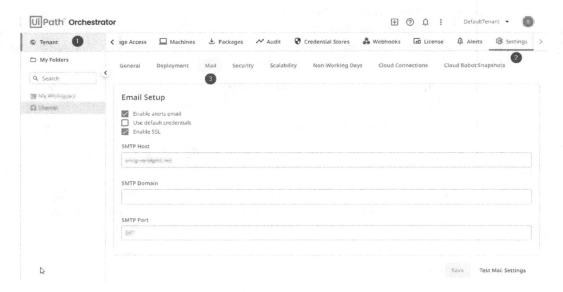

Figure 3.5 – Mail settings

II.   The next option is to set up a custom alert using an external monitoring tool. To set up that option, we need to enable the webhook on UiPath Orchestrator. Then the webhook will send faulted job events to an external monitoring tool (such as Splunk) events collector, and alerts can be set up to act on these faulted events. These alerts can be set up for critical processes to send text messages and email alerts to the on-call UiPath support team member.

> **Tip**
>
> It is recommended that the business-critical bot process has individual monitoring and alerts set up to guarantee support from the UiPath team. Suppose the RPA program is mature with more than 50 processes, saving $2 million in operational costs. Monitoring all the job summary emails will not be possible, so having a customized monitoring solution using external monitoring tool is highly recommended.

Infrastructure level

UiPath support and monitoring teams will need to monitor the infrastructure that supports the UiPath platform. Infrastructure resources can be physical or virtual, including workspaces, virtual desktops, databases, and servers. The infrastructure monitoring should cover all the UiPath components such as Orchestrator, **High Availability Addon (HAA)**, Robots, Test Manager, Insights, Kibana, and Databases. In most enterprises, a dedicated infrastructure support team is responsible for maintaining the health of these resources. Still, the UiPath support and monitoring team also needs visibility on the infrastructure health status:

- **Infrastructure monitoring dashboard**: It is recommended that a monitoring dashboard be configured for all the UiPath infrastructure resources in production and non-production environments. These dashboards can be hosted along with the enterprise monitoring application dashboards such as Splunk and BeyondTrust. The setup will be dependent on requirements received from the UiPath support team. The usual metrics measured are CPU, memory, and disk space in those infrastructures. Additional health metrics can be tracked in this dashboard, such as planned maintenance and outages.

- **Infrastructure failure alerts**: There are two primary options for alerts that can be set up for infrastructure monitoring:

  I. The first one deals with infrastructure error email alerts for faulted infrastructure resources. For instance, if an application server hosting UiPath Orchestrator or Database is down, an email alert will be sent to the respective teams. These alerts will be sent periodically until the resources are back to a healthy state.

  II. The second one deals with infrastructure warning email alerts. They are usually configured to warn the respective teams of unwarranted behavior in infrastructure resources such as VMs, servers, and databases. For instance, if there is a CPU spike in the database server, the overall UiPath platform performance will be reduced. This alert will be helpful for the respective team to mitigate the risk before it becomes an outage.

> Tip
> It is equally important to minimize false alerts. Otherwise, it will introduce compliance in the monitoring team, which will ultimately lead to alerts not being handled. This statement applies to all the three monitoring alerts discussed in this section.

I hope this was interesting! We will deep-dive into monitoring in action in a future chapter. Now, let's introduce a new concept for Jennifer and look at the details on the UiPath reporting setup in the next section.

UiPath reporting setup

UiPath RPA programs cannot sustain and grow without reporting. The scope of the reporting covers all the aspects of operations performed in the UiPath RPA CoE or program. Most reports can be generated from the existing monitoring setup for the UiPath CoE or the program.

The metrics that need to be measured and associated with respective reports in the UiPath CoE should be identified. The UiPath support and monitoring team will serve all the reporting requests.

In ABC Insurance Corporation, they use the UiPath Insights dashboard to create different dashboards, and reports can be scheduled from these dashboards. The best practice here is to have all types of metrics reported in dashboards so that you can always get the real-time statistics, and a snapshot of the dashboard is sent out periodically to stakeholders.

Let's try to understand by detailing the ABC Insurance Corporation's UiPath reporting framework:

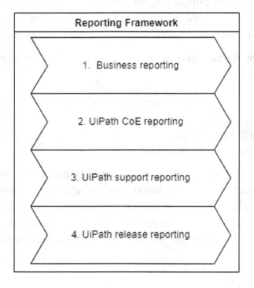

Figure 3.6 – ABC Insurance Corporation's UiPath reporting framework

Business reporting

The UiPath support and monitoring team is also responsible for generating reports for business stakeholders of the UiPath program. These reports are designed to communicate two primary pieces of information, which are usually the following:

1. Metrics (such as KPIs) relating to **return on investment** (**ROI**), cost savings, and human hours saved.

2. Reports of the health and risk levels of the current bot processes in production. These reports are categorized based on different business groups such as finance and the supply chain. The main objectives of these reports are to affirm business leadership of the value and confirm the funding to the RPA program.

Let's understand the different considerations for setting up these reports for business stakeholders:

*   **Bot-specific KPI report**: This report can be set up manually or automatically scheduled from the monitoring tool. As the monitoring dashboard tracks each bot's metrics such as ROI, cost, and time savings, custom queries can be set up to generate this report by tapping into monitoring log data in Orchestrator Database tables, custom logs tables, or Elasticsearch data. All the bot process metrics that belong to a particular business group are consolidated and reported together. They can be scheduled to deliver every month or quarter based on the business needs.

*   **Bot health report**: This report can be set up manually or automatically scheduled from the monitoring tool. Bot health metrics such as overall failure rates, downtimes, and SLA breaches will help calculate the overall health of the process and its risk ratings. The monitoring dashboard tracks these metrics; custom queries can be set up to generate this report by tapping into monitoring log data in Orchestrator Database tables, custom logs tables, or Elasticsearch data. All the bot health metrics that belong to a particular business group are consolidated and reported together. They can be scheduled to deliver every month or quarter based on the business needs.

> Tip
>
> These two reports are just sample reports shared with ABC Insurance Corporation's UiPath program business stakeholders, and there can be similar reports that are strategically important to business stakeholders. It is best to archive these reports in configuration management or knowledge management applications for future reference or during an audit.

UiPath RPA CoE reporting

Unlike business reporting, CoE reports are targeted to report on UiPath CoE performance, which includes the health of the process and platforms executing the CoE objectives. They are usually shared with the C-level executives and stakeholders who host and fund the UiPath CoE organization. Let's try to get some context on setting up these reports:

- **UiPath CoE-specific KPI report**: This report can be scheduled as per leadership requests. The overall picture of bot process metrics such as overall volumes and average handling time (per process) is covered by this report. Metrics such as delivery velocity, quality, value, cost and time savings, and the total cost of ownership by the UiPath CoE operation are configured in this report. The UiPath CoE maturity attributes covered in *Chapter 2* are also reported. They are usually scheduled to run every quarter to provide insights into the next quarter's planning. The main idea of this report is to give a deviant score of the bot process performance compared to the expected baseline metrics.

- **UiPath Platform health report**: This report can be automatically scheduled from the monitoring tool. These reports cover metrics such as process, platform, and product health metrics, including availability, downtimes, failure rates, and high-priority incidents. Custom queries can be set up to generate this report by tapping into monitoring log data in a monitoring tool, Orchestrator Database tables, custom logs tables, or Elasticsearch data. They are usually generated for all the UiPath products deployed in the production environment. They can be scheduled to deliver every month or quarter based on the business needs. The main idea of this report is to document the performance of the UiPath Platform infrastructure.

> **Tip**
> The UiPath CoE might need to review the CoE lead or manager before sharing them with the leadership. It is recommended to show some trends with graphs in these reports to make them more appealing.

UiPath support reporting

UiPath support reporting aims to report on UiPath CoE Support team performance to the UiPath leadership committee. Support reports include the health of the support process, resources, and platforms executing the UiPath support objectives. They are usually used to assess the problem areas in the current UiPath support organization's performance and improve them. A couple of these report setups are explained here:

- **UiPath process support health report**: This report can be automatically scheduled from the monitoring tool using data from applications used to manage the UiPath support processes. It reports on individual bot support health and risk levels based on incidents, availability, and change management. The data can be aggregated to report on the bot process support health at the function or department level. All the bot process metrics that belong to a particular business group are consolidated and reported together. It can be scheduled to deliver every week or month based on the leadership needs.

- **UiPath CoE support report**: This report can be scheduled to be delivered every month. It covers metrics that cover the high-priority incidents and overall support SLA compliance and highlights high-risk or the main maintenance bot processes. It also covers how the support team's continuous improvement initiatives improved the general support team performance. Data is usually mined from the **IT Service Management (ITSM)** databases such as ServiceNow and **Application Life Cycle Management (ALM)** tool databases such as Jira.

Tip

It is easy to set up these support reports in a highly digitally mature organization where we have a close association between ITSM, ALM, change management, and monitoring applications. The same reports will take time to set up if the data is spread across different systems.

UiPath release reporting

The main objective of UiPath release reports is to communicate the details of new or changed UiPath bot processes in production. They are usually shared during a release cycle and communicated to all the program stakeholders. There are two reports used in ABC Insurance Corporation. They are detailed as follows:

- **Release notes and deployment summary report**: This report can be set up manually or automatically scheduled from an ALM application (such as Jira) and change management tools (such as ServiceNow data). The report will be set up to share all the UiPath release information and confirm whether the deployments were completed successfully. The data can be aggregated to all releases or categorized into group releases based on the function or department level. It is common to consolidate these reports and share them every month or quarter with business and IT leadership teams.

- **Production validation summary report**: This report can be set up manually and scheduled after the production validation of the bot processes is completed by the respective business stakeholders. The report will be set up to share the statistics and information on successful and failed UiPath bot process validations. The data can be aggregated to all confirmations of releases or even categorized into group releases based on the function or department level. It is common to consolidate these reports and share them every month or quarter with business and IT leadership teams.

Tip

If release management is handled by an enterprise build and deployment management team, the UiPath support team can coordinate to set up these recommended release reports.

We will look at reporting in action in a later chapter. Next, let's introduce Jennifer to some of the crucial considerations for the UiPath deployment setup.

UiPath deployment setup

This next enabler is an exciting one that deals with the deployment of UiPath processes into different environments such as development, test, UAT, and production. Let's discuss how the ABC Insurance Corporation UiPath deployment started to happen and matured to the next level.

Manual deployment

ABC Insurance Corporation started its UiPath deployment just like any other UiPath customer. They manually deployed the UiPath NuGet packages generated from UiPath Studio into Orchestrator and set them up in different environments. Let's understand the high-level process before getting into the setup details:

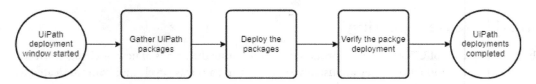

Figure 3.7 – ABC Insurance Corporation's UiPath manual deployment high-level process

- **Process**: The deployment process starts when the actual deployment timeframe starts. The deployment specialist will need a deployment plan and a package in place. Explicit instructions must be specified on the deployment details such as the process name, package location, destination folder to deploy, and deployment attributes such as dedicated robots, assets, triggers, and queues. Based on the details provided, the actual deployment of the package (along with all the customization requested in the deployment plan) will happen in UiPath Orchestrator. Once the packages are deployed, then validation of the deployed packages is carried out in Orchestrator. Usually, an email notification is sent to all the associated stakeholders.

- **Setup**: The deployment process needs to be documented and signed off by the UiPath leadership committee. Once the process is finalized, a pilot run needs to be planned in the available environment, and training needs to be provided to the deployment team. When a few test deployments are completed, the production deployment calendar must be updated and associated with the release management schedule. All the deployments should have an audit trail, and a dedicated team to support issues with the deployment should be available. Detailed deployment details and related topics are discussed in *Chapter 6*.

> **Tip**
>
> Manual deployments need to be controlled and restricted with role definitions from UiPath Orchestrator. Only a specific set of UiPath administrators should have access to deploy packages in Orchestrator. This is one of the common mistakes and is usually caught in an internal or external audit. Audit exceptions can derail your entire program, let alone the fact that they signify serious weaknesses in your program and can mean real risk to the whole governance process of deployment.

Automated deployment

Once the ABC Insurance Corporation streamlined their UiPath manual deployment, they moved into an advanced automated deployment setup. They were deploying the UiPath NuGet packages generated from UiPath Studio, using the Jenkins pipeline and the UiPath Jenkins plugin, into Orchestrator set up in different environments. Let's understand the high-level process before getting into the setup details:

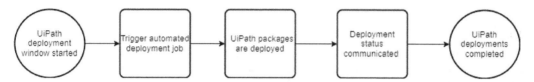

Figure 3.8 – ABC Insurance Corporation's UiPath automated deployment high-level process

- **Process**: The deployment process starts when the actual deployment time frame starts. The automated deployment pipeline will be triggered. The script will look for the deployment rules and package in the central repository. Deployment rules or configuration must be specified on the deployment details such as the process name, package location, destination folder to deploy, and deployment attributes such as dedicated robots, assets, triggers, and queues.

  Once all the artifacts and rules are checked, the actual deployment of the package happens in Orchestrator. An automated email notification is sent to all the associated stakeholders, and validation of the deployed packages is carried out in Orchestrator before the change tickets are closed.

- **Setup**: The deployment team must understand how the automated deployment of the Jenkins plugin works. Once the initial study is complete, they need to pilot in a sandbox environment. Once the Jenkins pipeline is confirmed, a test deployment needs to be completed with the help of this pipeline. Then, a common repository for storing the deployment artifacts needs to be finalized in applications such as JFrog Artifactory, along with the triggers for the Jenkins pipeline. Once the initial setup is finished, a pilot run needs to be planned in the test environment, and the team needs to provide training. When a few test deployments are completed, then a test production deployment should be triggered by this pipeline. Once the trial production deployment is successfully validated, we can produce the setup.

This deployment setup will automatically have an audit trail, integration to configuration management tools, and repositories such as GitHub. It is recommended to have a dedicated team to support issues with the automated deployment.

Visit this site for more information: `https://plugins.jenkins.io/uipath-automation-package/`.

> **Tip**
> Once a UiPath CoE reaches high maturity and has multiple deployments in a frequent release cycle, then automating the deployment will be necessary. The automated deployment principles and tools should align with the enterprise release management standards and policies.

We will detail the automated deployment through the DevOps concept in *Chapter 6*. Next, let's introduce Jennifer to some of the critical considerations for continuous improvement initiatives setup.

Continuous improvement setup for UiPath support

Continuous improvement initiatives are one of the most neglected UiPath support enablers, but they will be the most valuable to the entire UiPath program when implemented correctly. Ongoing improvement initiatives can be introduced right from the inception of the UiPath support team.

The ABC Insurance Corporation UiPath CoE support team used to get frequent incidents on a critical financial process that failed every day due to an application timeout exception. The Support team member would get involved to clear the error message, update the inputs sheet, and then rerun the job. As this was very frequent, it took a lot of time, and Jennifer identified this pain point and came up with an improvement plan to create a support utility that will perform these manual steps when this job faults. The support job will be triggered when the parent job fails. The job will perform the support steps and rerun the original job. The same approach was extended to other critical processes as well.

The support lead wanted to generalize the continuous improvements, so a framework was developed. Let's start with the continuous improvement framework implemented at ABC Insurance Corporation:

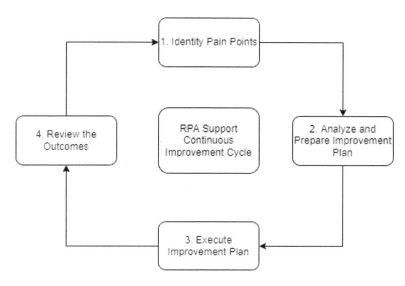

Figure 3.9 – ABC Insurance Corporation's UiPath support continuous improvement

- **Identify pain points**: The UiPath support and monitoring team will experience issues with the process, policies, SLA, or stakeholders. When they start to happen frequently, these issues become a problem, and when the root-cause analysis is made, a pattern emerges that will identify the pain points and bottlenecks.

- **Analyze and prepare an improvement plan**: Once the pain points are identified, all the relevant data points must be gathered to complete the analysis. This analysis will lead to the preparation of the improvement plan.

- **Execute improvement plan**: Once the improvement plan is approved and resources are allocated, the actual execution can happen with the help of internal and external team members. The implementation needs to track against the plan till the objectives are met.

- **Review the outcomes**: Once the plan is executed, the objectives of the improvement initiatives are reviewed by the UiPath leadership team to validate the results. Ideally, these outcomes will improve different KPIs, and it might take some time to realize the benefits.

Tip

Executing the improvement plans also need resources, so these initiatives cannot succeed without the support of the UiPath leadership team.

Continuous improvements for UiPath business KPIs

The improvement scope discussed here is initiatives that impact the UiPath products, bot process, and infrastructure. These continuous improvement initiatives will also involve changes to the bot process.

The setup consists of having a regular review of the bot monitoring metrics and reports. A dedicated team should be assembled to look for problem patterns from these reports and build an issue log with priorities. Then, support data needs to be compiled and presented to the UiPath process stakeholders to authorize the improvement initiative. It is recommended to have an ALM tool such as Jira to track project execution, and results must be communicated.

Once this setup is in place, they can be associated with UiPath program backlogs and the change management process.

> **Tip**
> Integrating the backlog and approval process is critical for resource allocation to these continuous improvement initiatives.

Continuous improvements for UiPath support operation KPIs

Unlike the previous initiative, the proposed improvement scope impacts the internal UiPath support organization. These continuous improvement initiatives will also involve changes to the processes and tools used to service UiPath support and monitoring requests.

The setup consists of having a regular review of the UiPath support organization performance metrics and reports. The existing UiPath support team should look for problem patterns from these reports and build an issue log with priorities. Then, support data needs to be compiled and presented to the UiPath support manager to authorize the improvement initiative. It is recommended to have an ALM tool such as Jira to track project execution, and results must be communicated. These internal initiatives can be executed without the need for extra resources in many cases, and have a faster cycle time for completion as well.

Having this setup will be a game changer in motivating the UiPath support team, as it will improve the technical and functional competency of the UiPath support and monitoring team.

> **Tip**
> Automating internal UiPath support and monitoring processes will benefit the UiPath CoE by improving compliance to all agreed metrics such as SLAs and cost savings.

Next, let's explain to Jennifer how to set up knowledge management for the UiPath support organization.

Knowledge management setup for UiPath support

A good knowledge management setup is mandatory for any successful UiPath RPA program. The UiPath support knowledge management strategy would be a subset and covered under a broader UiPath ROA CoE or program knowledge management strategy.

Sound knowledge management strategy and policies should be in place before setting up a knowledge management practice for the UiPath RPA CoE. A centralized knowledge repository that can scale up with the RPA CoE should be built, and access needs to be provided to the relevant stakeholders. Access can be delivered to internal UiPath CoE members, and restricted access can also be provided to other UiPath RPA stakeholders such as business, IT, and vendor teams.

Once the prerequisites are met, the main catalog for categorizing the UiPath RPA artifacts needs to be agreed upon. Then comes the actual knowledge upload, download, and administration activities.

> **Tip**
> The knowledge repository needs to be periodically curated, and stale or outdated artifacts must be archived to ensure relevance to the UiPath CoE team members.

The ABC Insurance Corporation RPA CoE's knowledge management pillars setup is detailed in the next section. Let's see the different steps that the RPA CoE took to mature their knowledge management practice:

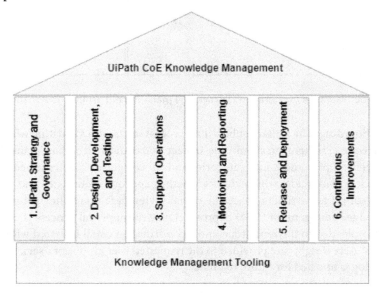

Figure 3.10 – ABC Insurance Corporation's UiPath knowledge management

Knowledge management pillars

In the ABC Insurance Corporation UiPath RPA CoE, there are six core categories and multiple subcategories for cataloging UiPath RPA CoE artifacts. The **Atlassian Confluence** application powered the knowledge management. The knowledge management journey started when the idea of the UiPath CoE was incepted. The order of categories was also carefully chosen to mature one class at a time based on the UiPath CoE needs.

The UiPath CoE Support manager was also given an additional responsibility to play a knowledge management leadership role. The knowledge management policy and scope of artifacts and restrictions must be laid out and signed off. For instance, the UiPath CoE can only upload documents, images, and recordings, but credential details are not maintained in these repositories. Once the policies are finalized, then the setup of the knowledge management starts with defining the governance process to onboard, retain, and retire artifacts.

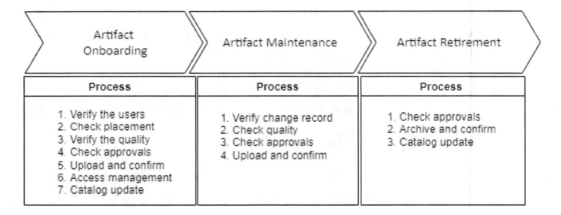

Figure 3.11 – ABC Insurance Corporation's knowledge management artifact governance process

- **Artifact onboarding**: The artifact onboarding process starts with validating whether the user has the correct access rights to submit the artifact to the knowledge management repository. Then, the relevant placement into categories and subcategories is determined. A high-level quality check is also made, which checks whether approved templates and standards are used in the submitted artifacts. Once the initial reviews are done, they get introduced for the **Knowledge Management (KM)** approval. Once the approval process is completed, the artifact gets uploaded to the repository, and a confirmation email is shared with the relevant stakeholders. Access rights are provided to the requested user groups or users, and finally, the artifact catalog is updated for future reference.

- **Artifact maintenance**: This is an ongoing process, which starts when an artifact is updated, and a change record is created. Once the request is received, the high-level quality checks are performed, and the document upload happens only when the knowledge management manager approves the update. Once the new version is uploaded, a confirmation email is shared to stakeholders who have access privileges to this artifact.

- **Artifact retirement**: The artifact retirement process is triggered when the artifact is no longer relevant to the UiPath CoE. The KM manager will verify artifact retirement requests. Once the request is approved, the artifact is archived, and a confirmation email is sent to the relevant stakeholders who had access privileges to this artifact.

It is vital to frequently update the artifacts with the latest information gathered from support and monitoring exercises. The artifacts' governance is a shared responsibility of the RPA CoE members, and the KM manager is usually an auditor of this governance process. The benefits of having a sound knowledge management setup from the start of the UiPath CoE will be realized once the artifacts are regularly onboarded and maintained. For instance, the ABC UiPath support team realized that when there was a high-priority incident, the team needed to respond in a time-sensitive manner.

We will look deeper into various artifacts included in the knowledge management setup in a different chapter.

> Tip
>
> The categories mentioned in this section are just references, and each UiPath CoE can customize the knowledge management strategy to fit its needs.

Tooling setup

In the ABC Insurance Corporation UiPath RPA CoE, Atlassian Confluence is used to set up the UiPath RPA CoE knowledge management. There are different options, such as Google Docs, Microsoft Teams, and SharePoint, to consider as well.

It is good to adopt the existing enterprise knowledge management tool to avoid setup and training costs. The device must have configuration management and disaster recovery to minimize risk:

- **Training**: Tool administration training needs to be provided to the UiPath support team that is responsible for maintaining the artifacts in the KM repository. Training must include adding and removing subcategories or folders, user access management, reporting, and configuration management options.

- **Access management**: An access management guidance policy needs to be established. This policy should dictate policies such as the following:

  A. User group with **Create, Read, Update, and Delete (CRUD)** privileges.

  B. The kind of access right a user receives.

  C. An approval process should be established to request access rights. Depending on the established guidelines, access rights can also be provided to a folder, subfolder, and artifact level.

> **Tip**
> Having a dedicated team to support the knowledge management application is necessary to help maintain the growing UiPath CoE or program.

We have walked through all the sections of this chapter; it's time to recap what we have learned in the following section and summarize the training concepts for Jennifer.

Summary

UiPath Support enablers' setup is the commencement event to mature an UiPath CoE and scale its support operations. The importance of the enablers and their setup details are the main objectives this chapter conveys to you in aspiring to work in UiPath support and administration roles.

We identified and fitted the enablers in the UiPath support framework and then introduced the importance of each enabler to the overall UiPath CoE and support operation.

This was followed by introducing the monitoring framework and explaining how to set up the various monitoring layers such as business, application, and infrastructure level monitoring. Then, the reporting framework was detailed, and specific reports set up in each of the reporting categories were discussed. We then understood the deployment setup, covering manual and automated processes and details. The next section covered the continuous improvement setup for overall UiPath COE delivery and UiPath support operations.

Finally, we covered the knowledge management pillars and processes to onboard, maintain, and retire, and the importance of having a tooling setup.

We used the ABC Insurance Corporation and Jennifer's persona throughout this chapter to ensure continuity as we did with the previous chapter. We will also use the same persona's training experience to explain the rest of the chapters.

The previous chapter and this one were designed to be theoretical, as many UiPath support and monitoring professionals lack the foundation to build a sustainable UiPath CoE. Now that this is out of the way, let's discuss more practical concepts in the future chapters.

This concludes all the chapters in *Part I, UiPath Platform and Support Setup*. In the next chapter, we will move to the next part, *Part II, Administration, Support, DevOps, and Monitoring in Action*, starting with UiPath Orchestrator administration.

Part 2: UiPath Administration, Support, DevOps, and Monitoring in Action

In this part, you will get an overview of UiPath Orchestrator components in detail, learn the best practices in the administration and configuration of UiPath Orchestrator, discover UiPath robot management and support details, learn the best practices in handling common support categories, and understand DevOps principles. In addition to this, you will also learn how to set up a UiPath Jenkins pipeline and integrate it with enterprise change management, and finally, you will get an overview of RPA monitoring framework components and monitoring tools and options.

This section contains the following chapters:

- *Chapter 4, UiPath Orchestrator Administration*
- *Chapter 5, Robot Management and Common Support Activities*
- *Chapter 6, DevOps in UiPath*
- *Chapter 7, Monitoring and Reporting in UiPath*

4

UiPath Orchestrator Administration

UiPath Orchestrator is often described as the heart of the UiPath platform, and hence UiPath Orchestrator administration becomes a core responsibility for UiPath Support personnel. This chapter will start by providing an overview of UiPath Orchestrator. You will understand how to perform UiPath Cloud administration tasks, and use UiPath Orchestrator to perform tenant- and folder-level entity administration. Finally, we will end with orchestrator uses cases and best practices.

A significant amount of time in a day is spent by UiPath Support personnel on administering UiPath Orchestrator operations in a production environment. We will extend Jennifer's induction into ABC Insurance Corporation's UiPath team as a use case to explain the chapter contents. Jennifer just concluded her UiPath support and monitoring training. She is ready to work on support requests.

> **Note**
> UiPath Orchestrator administration is one of UiPath's most critical components and where many UiPath support interview questions are usually directed. Hence, this is an essential chapter for any new UiPath student.

Here is what you will do in this fourth chapter:

- Understand the UiPath Orchestrator components in detail

- Know the best practices in the administration and configuration of the UiPath cloud

- Learn how to manage all the tenant-level entities, such as robots, machines, and licenses

- Learn how to manage all the folder-level entities, such as automation, assets, and queues

- Understand different Orchestrator support use cases and best practices

Before explaining each feature, let's provide an overview of UiPath Orchestrator to Jennifer.

A UiPath Orchestrator overview

UiPath Orchestrator is the core product of the UiPath Platform, and it is used to manage robots, automation, and associated components. It acts as a control room for publishing automation packages, provisioning machines and robots, and running and monitoring automation jobs. It is mainly divided into two main feature sets of entities:

- Tenant level
- Folder level

As defined in *Chapter 1*, the **Tenant** level is the highest level of grouping of Orchestrator resources. Robots, machines, and packages are some of the entities used across a tenant. In ABC Insurance Corporation's UiPath Orchestrator, there are multiple tenants, such as finance, IT, and HR.

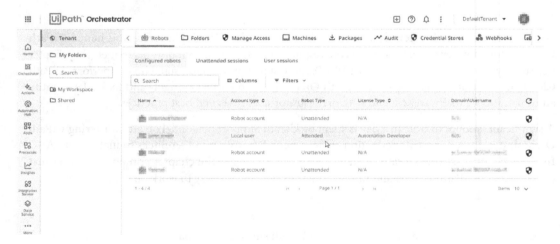

Figure 4.1 – Orchestrator tenant entities

The next level of grouping is at the folder level, where processes, assets, and queues are created to be used in the folder context. High-level monitoring capabilities are also provided in Orchestrator using the monitoring feature. We will cover the Orchestrator monitoring feature in a later chapter dedicated to UiPath monitoring.

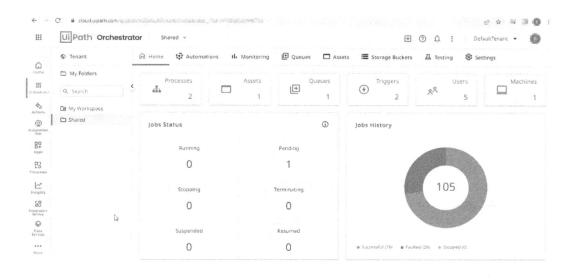

Figure 4.2 – Orchestrator folder entities

> **Note**
>
> UiPath support personnel should know all the entities and their association with the tenant and folder contexts. Only then they will be ready to address any UiPath production support issues quickly within the accepted SLAs.

Let's walk Jennifer through the Orchestrator terminology in the following section.

UiPath Orchestrator terminology

Before moving forward with the next section, new students need to understand the terminology used during UiPath Orchestrator administration:

- **Node**: This refers to the instances of UiPath Orchestrator that are deployed in different environments, such as a single-node or multi-node setup.

- **Tenant**: This refers to the first level of grouping of UiPath orchestrator resources. There can be multiple tenants in a single Orchestrator node, and they are divided based on functions in most UiPath programs, including those of ABC Insurance Corporation (such as finance and HR).

- **Folder**: This is used for the next level of grouping resources used in automation. They are usually divided based on business process functions in most UiPath programs, including ABC Insurance Corporation (such as IT service management and claims management).

- **Machines**: These are the host workstations (physical or virtual) for the robots to execute their automated processes. They are used for the next level of grouping resources used in automation. Machines are organized in multiple ways in UiPath programs; please refer to *Chapter 1* to refresh your memory. The machine template option is currently used by ABC Insurance Corporation.

- **Robots**: UiPath programs are executed on machines using robots. Unattended robots are used to execute processes where human interaction is not needed and are often scheduled with a trigger. Attended robots are used to execute a UiPath process by a human. There are both robots and attended assistants deployed in ABC Insurance Corporation.

- **Process**: This is the core UiPath deliverable, with the automation information deployed in Orchestrator. There are two primary types of processes: **foreground** (which need UI interactions) and **background** processes. Foreground processes are ones that interact with UI-based applications, and background processes are similar to services that run in the background with the UI interactions. Both types of processes are deployed in ABC Insurance Corporation's UiPath **Center of Excellence (CoE)**.

- **Jobs**: This refers to the actual execution of the UiPath process on the robot. Jobs move along different statuses such as the following:

  - **Pending**: When jobs are waiting for robot availability

  - **Running**: When the job is currently being executed by the robot

  - **Successful**: If the job is completed as expected

  - **Faulted**: When the job ended unexpectedly, and so on.

  Monitoring these job statuses is one of the primary tasks for UiPath support and monitoring personnel.

- **Triggers**: This refers to the event to initiate a job. The job can be scheduled to run based on a time-based trigger, or the queue item status update can initiate it. Both types of triggers are used in the ABC Insurance Corporation UiPath CoE.

- **Alerts**: This refers to real-time notification of recent action on robots, jobs, queues, triggers, and so on. They are actively used by ABC Insurance Corporation UiPath support personnel during production support activities.

- **Logs**: Here, this refers to the actual execution log messages generated by a job running on a robot. They are used to troubleshoot UiPath production issues by any organization, including ABC Insurance Corporation UiPath support.

- **Packages**: Here, this refers to the .nupkg file generated by UiPath Studio. It is the automation program file that can be published to UiPath Orchestrator to be used to create the processes.

- **Assets**: This refers to the shared or reusable variable or credentials that the UiPath process can use. Global or account-specific scopes of usage can also be set up as needed. Both types of assets are used by the ABC Insurance Corporation UiPath CoE.

- **Queues**: This refers to a container that can streamline incoming job requests. Different queues can be created based on the automation requirements. They are actively used by the ABC Insurance Corporation UiPath CoE for large volume robot jobs.

- **Transactions**: This refers to individual queue items being worked on by a robot. They have a life cycle that includes **new**, **in-progress**, **successful**, **failed**, and so on. The ABC Insurance Corporation UiPath CoE support personnel will also monitor the transactions for long-running automation jobs.

- **Storage bucket**: This refers to a container for physical storage of automaton artifacts such as input files, output audits, and supporting rules. They are actively used by the ABC Insurance Corporation UiPath CoE. This is an optional feature that can be enabled.

- **Credentials Stores**: This refers to entities that can securely store the credentials used in UiPath automations. By default, the UiPath Orchestrator database can support credentials storage itself, but eternal vendors such as CyberArk and Azure Key Vault are also supported.

- **Test Manager**: This is an external UiPath product that will enable the UiPath Test Suite. It will usually be deployed in test and development environments. Test jobs need test robot licenses to be executed; hence, separate licenses are needed to enable this feature. This product is used for system and regression testing by the ABC Insurance Corporation UiPath CoE.

- **Webhooks**: This refers to integration features that enable external products to subscribe to UiPath orchestrator events such as job creation and job faults. The Splunk receiver is configured to collect all job events in the ABC Insurance Corporation UiPath CoE monitoring setup.

- **Orchestrator APIs**: These are REST-enabled endpoints that can perform UiPath Orchestrator operations without logging into the application. They are used by the ABC Insurance Corporation UiPath CoE support team to build utilities that can automate a few UiPath Support operations such as stopping long-running jobs and starting faulted jobs automatically. We will investigate this in detail in a later chapter.

Note

These are some of the core terminology used during UiPath Orchestrator, but there are many more, and they will be covered in the later chapters and contexts.

Now that Jennifer is familiar with the terminology, let's look at the different components that make up UiPath Orchestrator's capability.

UiPath Orchestrator capability

UiPath Studio (Pro/X) is used to build the automation script, and they are deployed as packages in Orchestrator. The automation processes are created and configured to be executed on robots deployed in machines. Once the robot audit logs execute the jobs, notifications are produced by Orchestrator. Tenant-level and folder-level entities administration will provide all the supporting features for sustaining the automation.

All the data is stored in the centralized orchestrator database, and customized storage of secure credentials can also be done with the help of third-party credential stores. Monitoring and testing solutions can be configured with the orchestrators. Webhooks and Orchestrator APIs provide additional integration capabilities with external applications to improve the usage of UiPath Orchestrator.

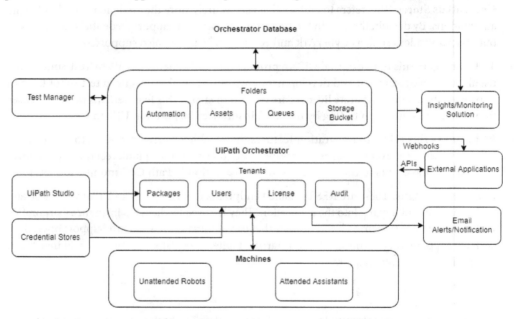

Figure 4.3 – UiPath Orchestrator Conceptual Diagram

The main capabilities of UiPath Orchestrator and its usage are detailed in this section:

- **Control and orchestration**: Managing and having centralized control of all entities and resources deployed in an automation environment and orchestrating automation that answers the what, when, and where questions are the two core capabilities of UiPath Orchestrator. Access to a 360-degree view in real-time information about automation entities will be a cherished capability for the UiPath Support team. In ABC Insurance Corporation, a UiPath support team uses this core capability to perform administration of the overall UiPath automation and measure the overall health of the UiPath platform.

- **User and account management**: Access management to UiPath Orchestrator (with roles and user access management) is another critical capability of UiPath Orchestrator. Roles can be defined for both user and robot accounts.

  There are two-step approval processes by the ABC Insurance Corporation RPA support team to manage user and account requests. First-level approval is given by the support leader and the next level will be done by a leader of the UiPath CoE committee. Limited UiPath support resources have access to add or edit roles.

- **Automation management**: Core UiPath automation is managed by UiPath Orchestrator. Managing automation artifacts such as packages, bots, machines, processes, jobs, and monitoring are some of the core features that support automation in UiPath Orchestrator.

  In the ABC Insurance Corporation UiPath CoE, RPA operation activities such as provisioning robots, deploying packages, executing jobs, and troubleshooting production issues with the help of logs are some of the core automation management activities where UiPath Orchestrator is heavily used.

- **License management**: The tenant-level licenses are managed by UiPath Orchestrator. **Unattended**, **Attended**, **Test**, **Developer**, and **Citizen** developer runtime licenses can be managed from a single centralized panel. As these licenses support the core automation, it becomes a core capability for the UiPath support team to constantly monitor.

  In ABC Insurance Corporation, a UiPath production and non-production product and runtime license usage are constantly monitored from UiPath Orchestrator. License expiry notifications are forwarded to the vendor management team promptly.

- **Integration management**: Integrating with other internal or external applications in the automation landscape is vital for any successful automation program.

  For instance, in ABC Insurance Corporation, UiPath production job events are shared to the Splunk monitoring tool receiver by the Webhooks feature. External applications such as Postman can create or stop UiPath jobs using the Orchestrator API. These integrations are possible in UiPath Orchestrator.

> **Note**
> In every new release of the Orchestrator version, Orchestrator capabilities are evolving with changes to technology and the business landscape. UiPath support personnel should be at the top of the learning curve and use all the available features to meet the request.

I hope Jennifer got the complete overview of UiPath Orchestrator; let's look at the details of each feature and its applicability in the next section. We will use **Scenario Tags** (**STs**) to highlight these real-life support requests.

UiPath cloud administration

The cloud administration is placed outside UiPath Orchestrator. However, UiPath support personnel must understand the features and have a strong understanding of the administration activities to be performed on them.

Orchestrator tenant administration

Only select users assigned to UiPath administration roles can perform these administration activities. Tenant management and its license allocation are the two main activities that can be performed.

Figure 4.4 – Orchestrator tenants

Tenant management

Management of tenants in this context refers to adding new tenants, disabling existing tenants, and deleting tenants. When adding a new tenant, the UiPath Orchestrator administrator must choose the services based on the license purchased at the tenant level.

ST1: In the ABC Insurance Corporation UiPath CoE, a new support request is for creating a new tenant for finance functions. A new strategic automation initiative is getting kicked off this year. Jennifer checks for all the required approval, and then she logs in to the UiPath cloud administration pane and fills in the required details, adds a new tenant called **Finance**, and adds Insights' service as requested in the support request.

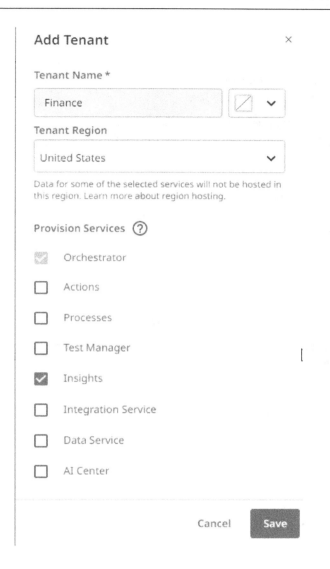

Figure 4.5 – Adding a new tenant

Once the changes are saved, the new tenant will be available, and the required licenses need to be configured.

Tenant license management

License for different services such as Orchestrator and Insights are allocated at the tenant level and these licenses are available for users assigned to these tenants. License for robots such as attended, unattended, and test are configured at this level.

ST2: In continuation from the previous scenario, Jennifer allocates five attended and unattended bot licenses to the newly added finance tenant as requested in the original request.

> **Note**
>
> In the ABC Insurance Corporation UiPath CoE, there are multiple tenants. The UiPath CoE leadership committee needs to approve any tenant-level management activity before changes are made at the tenant level.

User and group management

Users, robots accounts, and groups can be added at the UiPath administration level to have shared access to all the UiPath products, including Orchestrator.

ST3: Jennifer receives another request to add a new attendee user and an unattended robot. The user and robot account information is shared in the request.

Figure 4.6 – User accounts

To add a new attended user, Jennifer needs to use the **Invite Users** option and choose the **Automation User** group (which holds the five allocated attended licenses). Once the user chooses to complete the request, they will be provisioned to the tenant to execute attended automation. The same user also needs to be added at the tenant and folder levels to start executing the jobs. We will do this step when we reach the tenant entity management section.

To add an unattended robot, Jennifer must add a robot account and just leave the group as default. The actual group will be added to the tenant entity management section.

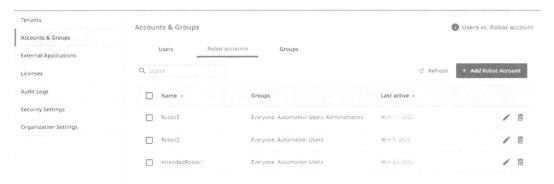

Figure 4.7 – Robot accounts

> **Note**
>
> In the case of ABC Insurance Corporation, they also had **Level 0 UiPath support**. Level 0 covers the RPA support service catalog that lists all the self-services available to end users, such as password reset or rerunning a failed job. As the support model matures and moves toward touchless mode, self-service catalogs are getting popular over time. A self-service catalog can also be extended to all levels of UiPath support as well.

New user group setup

The Orchestrator groups are used to bundle users with a similar business objective and access rights. This section displays the available groups; the user can view the allocation rules and add users or robots to the available groups. Whenever there is a new business or technical initiative, or changes to the existing organization, then the UiPath support team will receive requests to manage these groups.

ST4: Jennifer gets a new request to create a new group called **Business Users**, leading the citizen development initiative at ABC Insurance Corporation.

She creates the requested new group and adds the users. In addition to this, she allocates **Citizen Developer - Named User** licenses to this group by editing the allocation rules.

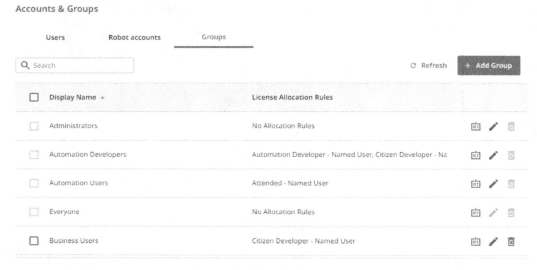

Figure 4.8 – Orchestrator groups

> **Note**
>
> In the ABC Insurance Corporation UiPath CoE, separate user groups are maintained for logical grouping of users, based on UiPath application licenses such as citizen developers and Insights users.

Next, let's look at some licenses and other options in the **UiPath cloud administration** tab.

License and other settings management

All the licenses purchased by the client organization can be viewed from the **Licenses** tab. There are two main sections:

- **Users**
- **Robot & Services**

User-level licenses

Jennifer can use this section to understand an existing license at the user level. The allocation rule for the newly created **Business Users** group can be viewed in this section. User allocation rules can be customized or inherited from groups as well.

Figure 4.9 – User-level licenses

Robot and Services licenses

Jennifer can view all the available robot licenses and active UiPath services, such as **Document Understanding**, **AI Robot**, and **Insights**, from this section. Tenant-level license allocation is also tabulated in this same tab.

Figure 4.10 – Robot and service-level licenses

Security settings

Single sign-on (SSO) can be activated from this section. The enterprise identity management team needs to be involved when activating SSO for the UiPath cloud.

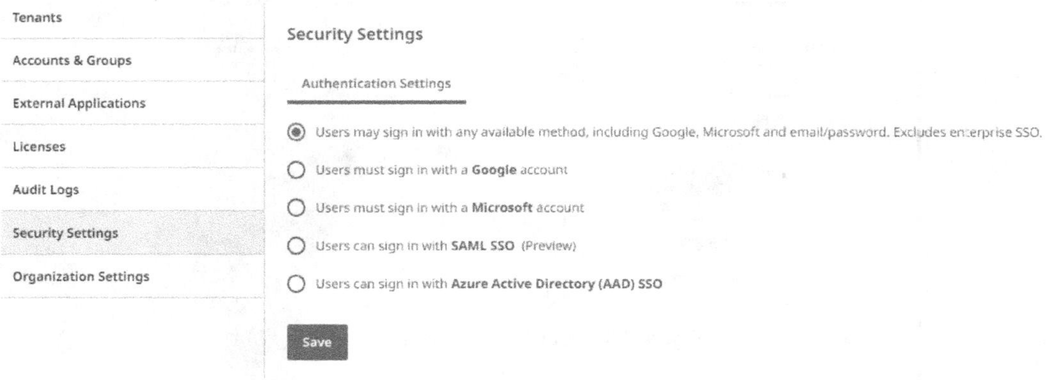

Figure 4.11 – Orchestrator security settings

Please refer to this link for more details: `https://docs.uipath.com/automation-cloud/docs/authentication-settings`.

Organization settings

The highest level of structuring for the UiPath Automation cloud can be updated by renaming the organization name. Another great feature available here is being able to update the language settings for an organization. A **Support ID** option for raising tickets with the UiPath global support team is also found here.

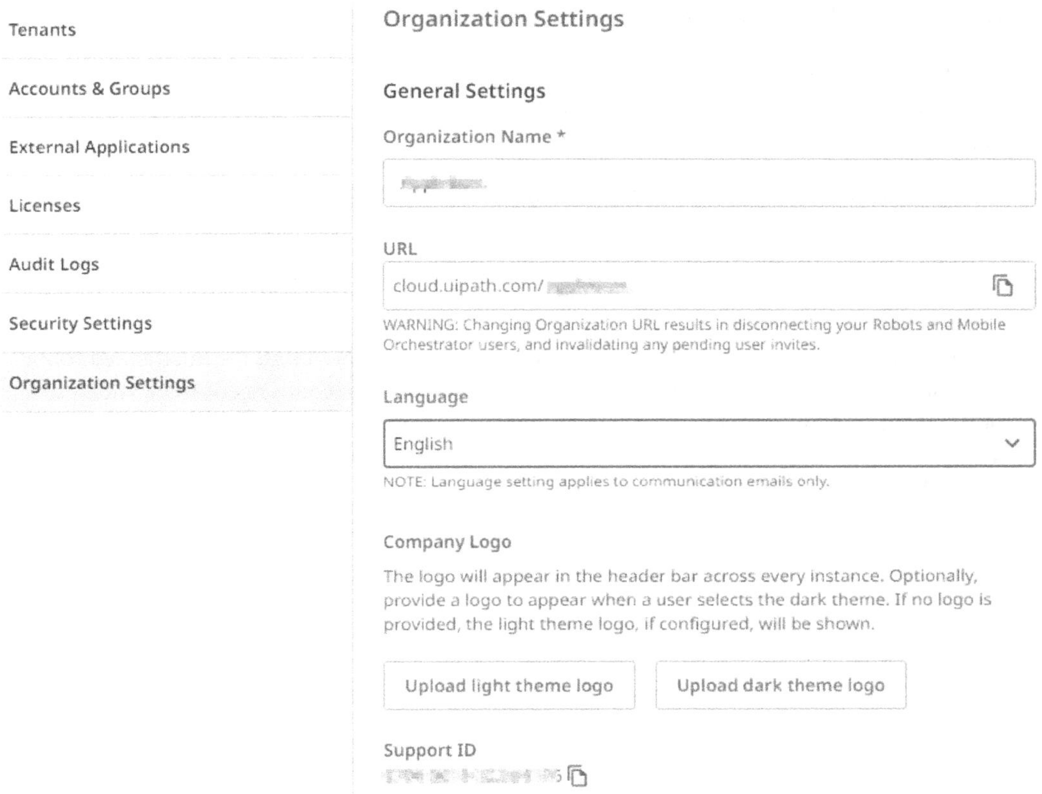

Tenants

Accounts & Groups

External Applications

Licenses

Audit Logs

Security Settings

Organization Settings

Organization Settings

General Settings

Organization Name *

URL

cloud.uipath.com/

WARNING: Changing Organization URL results in disconnecting your Robots and Mobile Orchestrator users, and invalidating any pending user invites.

Language

English

NOTE: Language setting applies to communication emails only.

Company Logo

The logo will appear in the header bar across every instance. Optionally, provide a logo to appear when a user selects the dark theme. If no logo is provided, the light theme logo, if configured, will be shown.

Upload light theme logo   Upload dark theme logo

Support ID

Figure 4.12 – Organization Settings

Note

In the ABC Insurance Corporation UiPath CoE, only the system administrators can update the security and organization settings, with proper approval from the IT and UiPath RPA program leadership teams.

I hope this was interesting for Jennifer. Now, let's look at the details of entities covered under the tenant level in the next section.

Tenant-level entities administration

Tenant-level entities administration can be performed once we log in to UiPath Orchestrator and choose the tenant where we need to work. Context and data of the tenant-level entities will change based on the tenant the users log in.

In the ABC Insurance Corporation UiPath Orchestrator, the **Default** tenant contains the IT-related automation resources, and Jennifer logs into that tenant to perform her tasks. Let's start with the robots.

Robots

This section will list all the robots available in the tenant. Jennifer cannot find the unattended robot account she has added at the cloud administration level. The robot account needs to be added to a folder for the robot to show up in this list. Once Jennifer adds the robot, it shows up in the configured robots list.

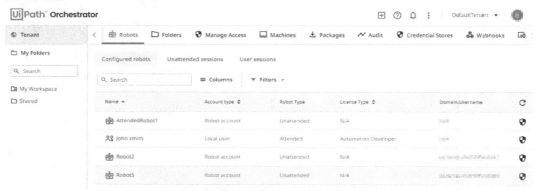

Figure 4.13 – Orchestrator robots

ST5: As part of Jennifer's monitoring jobs, she needs to check whether all the host machines are available in Orchestrator. The **Last Heartbeat** information can be found in **Unattended sessions**. She uses this feature frequently for her monitoring tasks.

Figure 4.14 – Unattended sessions

ST6: Jennifer also checks all users' statuses, including the newly added attended user she recently added from the **User sessions** tab.

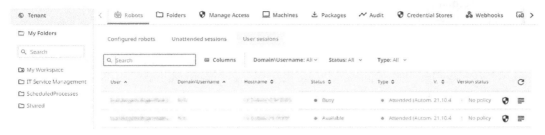

Figure 4.15 – Attended sessions

Folders

This section can be used to organize the automation based on various processes or any logical grouping.

ST7: Jennifer adds the new attended user and unattended robot account to this **Shared** folder and makes sure the correct machine templates are added to the respective folder.

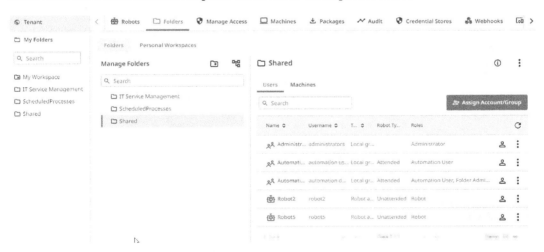

Figure 4.16 – Orchestrator folders

Managing access

This section can manage access by assigning existing roles to the users and robot accounts. Robots are often shared between different folders in case of volume surges, and the UiPath support team has to support these ad hoc requests from business or operation teams.

ST8: Jennifer gets a temporary request to remove **Robot5** for a day from a particular folder so that the robot can be utilized in a different one. She can perform her activity from this manage access page. Before she can delete **Robot5**, the robot account needs to be removed from the **Shared** folder first.

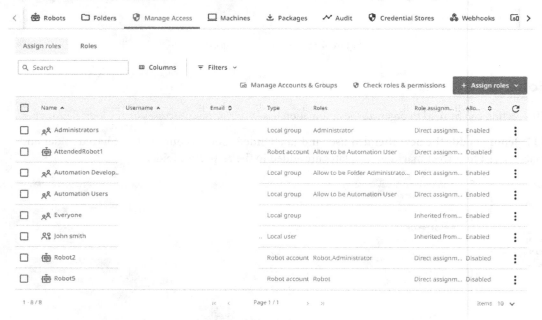

Figure 4.17 – Orchestrator access management

ST9: Jennifer receives a new request to add a new role for the tenant and folder levels. She needs to click + **Add a new role** to perform that task.

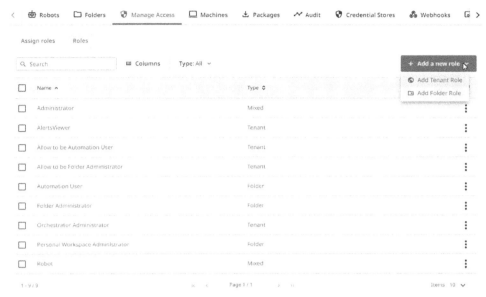

Figure 4.18 – Orchestrator roles

ST10: Business users need view-only access to available robots, machines, and alerts. Jennifer configures the requested role.

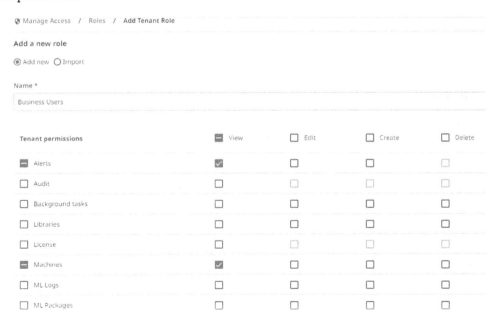

Figure 4.19 – Orchestrator tenant role

ST11: IT administrators need access at the folder level to view storage buckets and monitor long-running jobs. Jennifer configures the requested role.

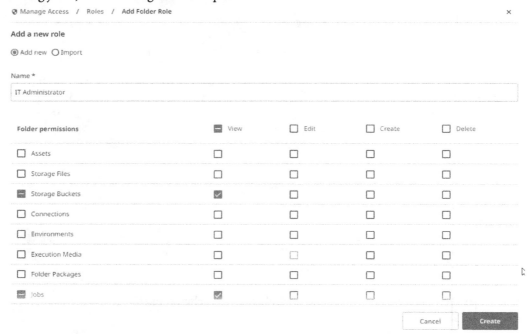

Figure 4.20 – Orchestrator folder role

Machines

The **Machines** section lists all the available machine options in UiPath automation.

ST12: Jennifer can check the machine template key from this section to connect the unattended robot to Orchestrator. Please refer to *Chapter 1* for all the details on machine types.

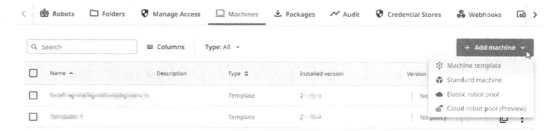

Figure 4.21 – Orchestrator machines

Packages

All the available packages deployed in the tenant are available here. UiPath Orchestrator administrators have access to upload new NuGet (`.nupkg`) packages as well.

ST13: There is a new version of the package pushed into the production environment, and Jennifer is tasked with upgrading this version to the published processes in production.

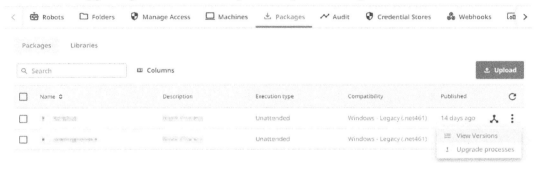

Figure 4.22 – Packages

Audit

This section will be helpful for UiPath support to find the sequence of activities when they are troubleshooting a production issue.

ST14: Jennifer recently restarted a job manually as requested by the business, and she can find the audit details from this section as well.

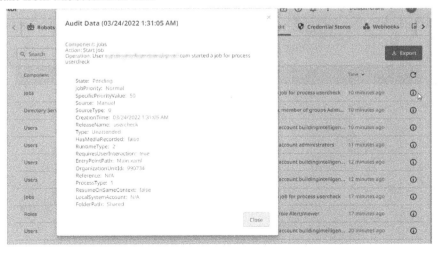

Figure 4.23 – Audit

Credential stores

In the ABC Insurance Corporation UiPath CoE, the default Orchestrator database stores the credentials.

ST15: As part of a finance process transformation program, there is a proof-of-concept request in the Test Orchestrator to enable and use Azure Key Vault. Jennifer needs to use this section, and with the help of the Azure administrator, she can configure the new credentials stores for the finance tenant.

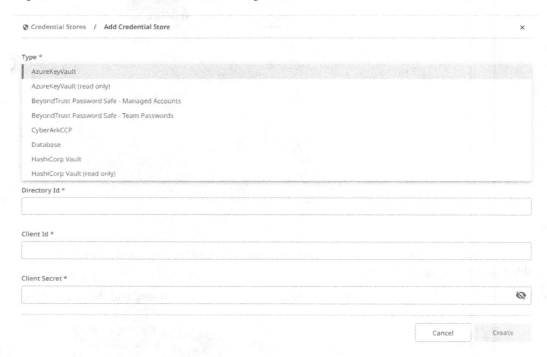

Figure 4.24 – Credential store options

Webhooks

An external Webhooks events receiver can be set up in this section, and the kind of Orchestrator events that need to be transmitted can also be configured.

ST16: Jennifer received a request to configure a new Webhook for all the jobs completed and faulted in the tenant. Hence, she configures the same and confirms that the events are sent to the collector before she closes the request.

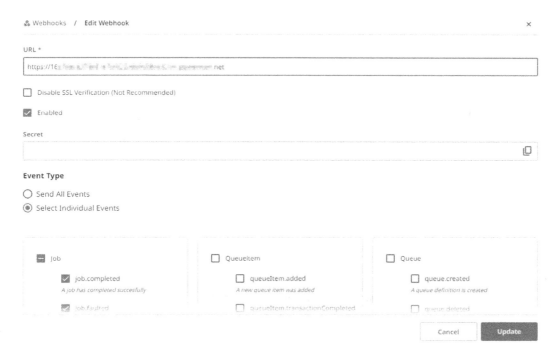

Figure 4.25 – Orchestrator Webhooks

License

This section can monitor all the tenant-level license allocations. It is one of the regular maintenance activities of the UiPath support team to check the available license utilization. Runtime licenses will decide the number of robots available in a particular machine template at a point in time, and they can be set up in this section.

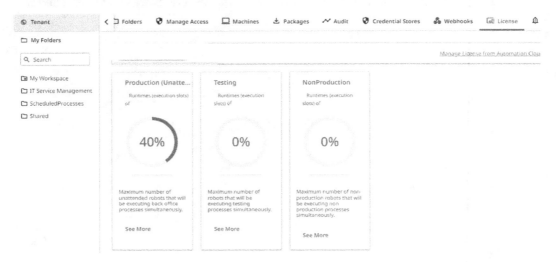

Figure 4.26 – Orchestrator licenses

ST17: As Jennifer added a new unattended robot as part of a support request, as part of this request, she needs to provide the runtime license to the machine template by navigating to the **See more** section.

Figure 4.27 – Runtime license

Alerts

This section has all the alerts generated by robots, transactions, triggers, queues, and so on. The support team can enable or disable types of alerts as well.

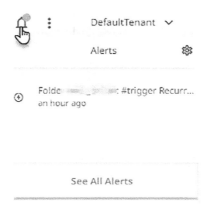

Figure 4.28 – Orchestrator alerts

ST18: During the UiPath production jobs monitoring and support window, Jennifer will actively look for faulted jobs alerts from this section to troubleshoot the issue.

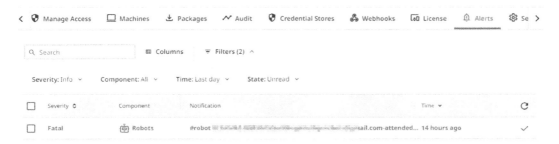

Figure 4.29 – Fatal robot alert

Settings – email setup

Email setup is usually configured during the UiPath Orchestrator setup, and it is mandatory to receive email alert notifications from Orchestrator to registered users.

ST19: Email alerts are not being delivered, and there is a support ticket assigned to Jennifer to troubleshoot. She logs into the **Email Setup** tab, uses the **Test Mail Settings** option, and confirms that email alerts are not being sent. Then, she discusses this with the email administration team and finds out the **Simple Mail Transfer Protocol (SMTP)** password for the UiPath Orchestrator administration account has expired. She updates the correct password and fixes the issue.

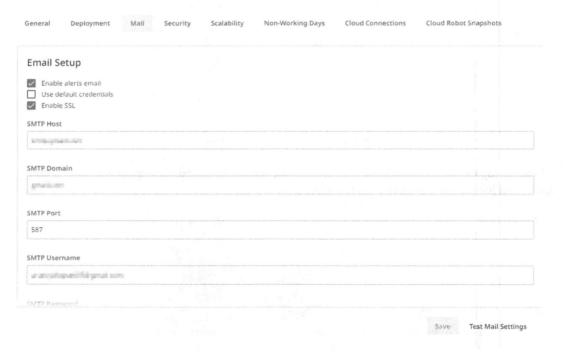

Figure 4.30 – Orchestrator mail setup

> **Note**
> This section only highlighted frequently used tenant-level entities, and there are many features that I encourage users to explore and learn from in the UiPath Orchestrator trial or community editions.

Next, let's take Jennifer through some practical assignments related to folder-level entity management in UiPath Orchestrator.

Folder level entities administration

Every folder in the tenant has a different context and each folder has unique automating data, such as processes, jobs, and assets, associated with them. It is crucial to have the folder-level administrator role perform a few administration tasks at this level in UiPath Orchestrator.

Lets take a look at few of the support requests in this section.

Automation

- **Process**: A UiPath process is created based on the packages fed to UiPath Orchestrator. This section lists out all the processes that are available in the folder. New processes can also be added based on the available packages and arguments (if applicable) and by providing additional settings, such as descriptions and job priority.

Figure 4.31 – Processes

ST20: Jennifer received a request to roll back a recently updated process in production, as it was continuously failing. She edits the target process and rollback to a previous stable version of the process. Then, she manually executes the same process to perform the test run.

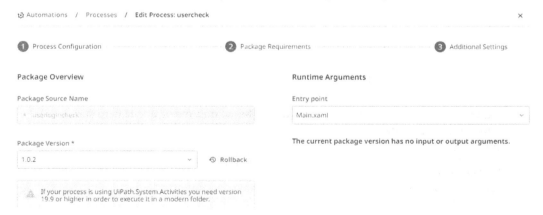

Figure 4.32 – Edit a process

- **Jobs**: The jobs section in Orchestrator lists all the jobs executed in the folder. This panel can view different job-related information, such as the process name, the robot name, type, and current state, which can be viewed from this panel. A new job can also be started from the same panel.

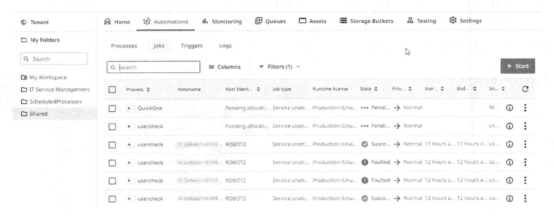

Figure 4.33 – Orchestrator jobs

ST21: Continuing from the previous scenario, once Jennifer ran the process manually, a related job was created, and she clicked on **Restart** to rerun the job and double-check before she could close her support ticket.

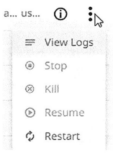

Figure 4.34 – The job context menu

- **Triggers**: Triggers are used to schedule the jobs based on time or queue item status. All the existing triggers are listed in this panel, and new triggers can also be added, or existing triggers can be edited from this tab.

ST22: Once the process was rolled back and tested, Jennifer got another request to enable the trigger on that process and update the run interval to 10 minutes. Jennifer edits the existing trigger and updates the interval as requested.

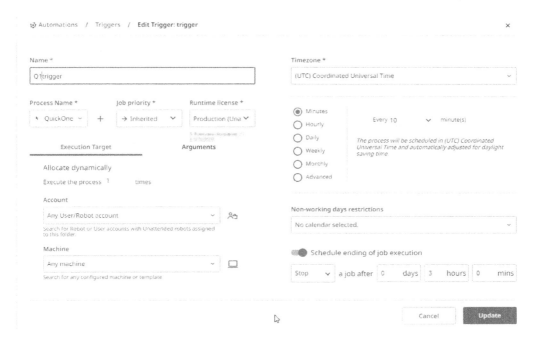

Figure 4.35 – Orchestrator triggers

- **Logs**: The **Logs** tab in the Jobs panel lists all the logs generated by all the executed jobs in that folder. They are vital for any troubleshooting tasks with the UiPath Support team.

  ST23: Once the process was triggered after 10 minutes, Jennifer checked the logs of that job to confirm a successful run.

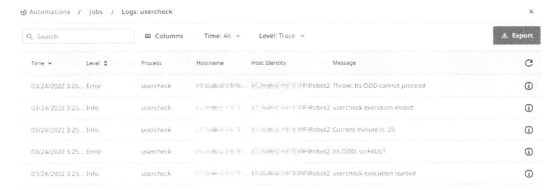

Figure 4.36 – Orchestrator logs

Queues

A list of all the queues in the folder is listed in this section, which will provide all the statistics of how many transactions are in progress, completed, exceptions, average time, and so on.

ST24: A new UiPath bot was developed and a change request **CRXXXX** was created to deploy this bot into production. Jennifer receives a support task as part of a change request (**CRXXXX**) to add a new queue for a new process. The incoming volume for this process is high, and multiple bots will be deployed to execute this process after the next release of the automation. She creates a queue in Orchestrator and, based on agreed CoE default values, adds an auto-retry and sets the SLA for this process as 1 hour.

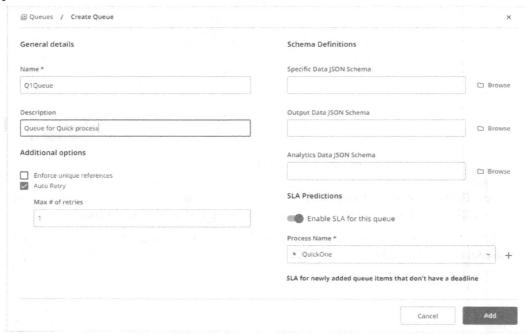

Figure 4.37 – Orchestrator queues

Assets

The **Assets** section lists all the assets available for use in this folder. Text, Boolean, integer, and credential types can be added as assets, and they can be scoped as global or robot-specific values.

ST25: Jennifer receives a support request to add a new **Asset** that stores the CRM application credentials. The credential details are passed in a secure file. Jennifer sets up a global value asset with the type set to **Credential** and updates the support task to a completed state.

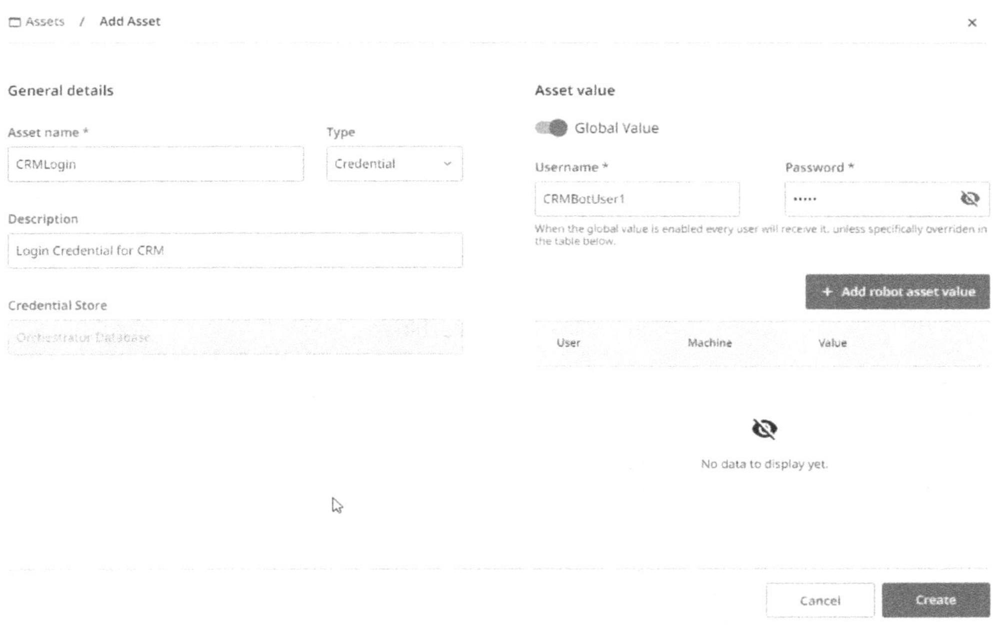

Assets / Add Asset
General details
Asset name *
CRMLogin
Type
Credential
Description
Login Credential for CRM
Credential Store
Orchestrator Database
Asset value
Global Value
Username *
CRMBotUser1
Password *
When the global value is enabled every user will receive it, unless specifically overriden in the table below.
+ Add robot asset value
User
Machine
Value
No data to display yet.
Cancel
Create

Figure 4.38 – Orchestrator assets

Settings

User and machine access to folders can also be viewed and edited from the **Settings** tab. In a few cases, UiPath support administration is restricted to the folder level; hence, this option becomes vital in those scenarios.

Tenant
My Folders
Search
My Workspace
IT Service Management
ScheduledProcesses
Shared
Home
Automations
Monitoring
Queues
Assets
Storage Buckets
Testing
Settings
Manage Access
Machines
Search
Assign Account/Group
Name
Username
Type
Robot Type
Roles
Administrators
administrators
Local group
Administrator
Automation Users
automation users
Local group
Attended
Automation User
Automation Develop...
automation developers
Local group
Attended
Automation User, Folder Administrator
Robot2
robot2
Robot account
Unattended
Robot
Robot5
robot5
Robot account
Unattended
Robot

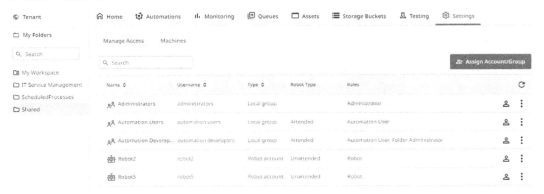

Figure 4.39 – Folder users

ST26: Jennifer receives a support request to remove the robot and machine template from the shared folder. She uses the **unassign** option to perform the task from machines tab and update the ticket status.

Figure 4.40 – Folder machines

> **Note**
> Other contexts, such as action, storage buckets, and testing, are not covered in this section. As this is an evolving area, UiPath Support personnel should constantly look for updates from UiPath releases.

I hope Jennifer was able to take in and use all the folder context learnings in the preceding assignments. Let's move on to the next section of the chapter.

UiPath Orchestrator use cases and best practices

This last section of the chapter explains common use cases and best practices to Jennifer. There are multiple ways UiPath Orchestrator can be used in a UiPath CoE. The most common use cases from ABC Insurance Corporation are detailed in this section. Understanding these use cases will provide an edge for UiPath Support personnel such as Jennifer during actual UiPath Orchestrator administration activities.

Use cases

The use cases listed here are formulated based on the type of robot connected to UiPath Orchestrator in the production and non-production environments of the ABC Insurance Corporation UiPath CoE. Having a good understanding of these existing use cases will give Jennifer the confidence to work on support issues.

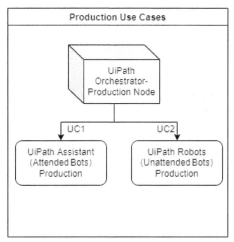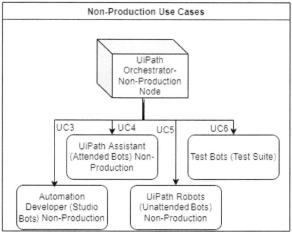

Figure 4.41 – ABC Insurance Corporation Orchestrator use cases

Lets try to understand various use case details in this section.

- **UC1 – production attended**: The UiPath Orchestrator production node is connected to UiPath assistants, and users can run attended automation on them on demand.

- **UC2 – production unattended**: A UiPath Orchestrator production node is connected to UiPath robots, and jobs can run in an unattended mode based on trigger conditions.

- **UC3 – non-production developer**: A UiPath Orchestrator non-production node is connected to UiPath Studio, and UiPath developers can develop and run attended automation to test the scripts on demand.

- **UC4 – non-production attended**: A UiPath Orchestrator non-production node is connected to UiPath assistants, and users can run attended automation on them on demand, usually for testing attended automation.

- **UC5 – non-production unattended**: A UiPath Orchestrator non-production node is connected to UiPath assistants, and test jobs can run in unattended mode based on a trigger condition.

- **UC6 – non-production test**: A UiPath Orchestrator non-production node is connected to UiPath testing robots, and test automation jobs can run from Orchestrator. This use case is only valid if the Test Suite licenses are available.

> **Note**
> There may be other Orchestrator use cases such as how they are used and products such as Document Understanding and AI Center. The same principle applies to configuring UiPath Orchestrator for new use cases.

Now that we have covered the use cases, let's walk Jennifer through some of the best practices followed by ABC Insurance Corporation and the UiPath Support team.

Best practices

UiPath Orchestrator's best practices and guidelines are already outlined in ABC Insurance Corporation's UiPath RPA CoE governance documents. These best practices need to be frequently updated based on support team feedback. Let's look at them in detail:

- **Naming convention**: All the objects created under Orchestrator should have proper naming conventions, starting from tenants, folders, processes, robot users, machines, assets, queues, storage buckets, and testing. Maintenance of Orchestrator resources will be simplified if naming convention checks are in place at the start of the UiPath operation.

- **Roles and user management**: Role definitions and user mapping to the roles need to be defined at the UiPath CoE level. This feature is one of the most audited ones; hence, having proper approvals and an audit trail in place becomes mandatory for the UiPath support team.

- **Tenant and folder management**: It is good to replicate production tenant and folder structure across different environments such as **Test** and **User Acceptance Testing (UAT)**. The automation processes should assign priority and business criticality indexes to the tenants and folders. This information will help to plan UiPath Orchestrator support and monitoring activities.

- **License management**: License management at the UiPath cloud level, tenant level, and folder level need to be managed with proper audit trails and approvals. License allocation needs to be in sync with the resources being used and needs to be frequently updated to maximize the value of licenses.

- **Package management**: Package management that includes upload, delete, and version update permission should be restricted. An audit trail of change management, application life cycle, and configuration management should be associated with every production package deployment.

- **Governance setup (create, read, update, and delete)**: A governance process should be in place for UiPath Orchestrator administrators to set up **Create, Read, Update, and Delete (CRUD)** permissions for the following Orchestrator entities:

  - **Roles**

  - **User**

  - **Folder**

  - **Machines**

  - **Assets**

  - **Queue**

- **Webhooks**

- **Credentials**

- **Central repository management**: It is good to maintain a central repository of all UiPath Orchestrators in a document or a specialized enterprise architecture modeling tool. For instance, having all the UiPath processes and their associated resources, such as assets, queues, business stakeholders, the SLA, business criticality, and risk information, will become very handy when troubleshooting production issues. A UiPath support lead should take accountability to maintain this artifact, and the information should be audited as well.

- **Cleanup of unused objects**: A UiPath Orchestrator entity's cleanup activities need to be planned and executed as per the interval agreed by the UiPath support leader. Allowing stale or unused objects into the repository may lead to UiPath production support issues.

> **Note**
> Many UiPath support teams often overlook best practices when they start, but having them in place will ultimately help the overall UiPath CoE and support team.

We have walked through all the sections of this chapter, and it's time to recap the learnings in the following section.

Summary

UiPath Orchestrator administration is one of the critical functions performed by UiPath support personnel regularly. Hence, it is vital to understand the technical aspect of its features and have a practical understanding of its applicability.

We started to walk Jennifer through an overview of UiPath Orchestrator, with a quick introduction of the entities and the user interface. Then, she understood all the key terminology used in UiPath Orchestrator before learning how these terms work together to form a conceptual view of Orchestrator.

In the next section, the core capabilities of Orchestrator were explained. Understanding how to use these capabilities during UiPath support activities formed the content for the rest of the chapter.

The practical scenarios were used to explain the later section of the chapter, starting with the core features of UiPath cloud administration, where an administrator can perform different tasks, such as tenant, user, and license management at a UiPath cloud product level.

We then introduced to Jennifer the administration tasks related to tenant-level entities, such as robots, machines, and packages. She was given different real-life tasks to perform to learn on the job.

The following section detailed the folder-level administration tasks related to tenant-level entities such as processes, jobs, and assets. Jennifer was able to perform the tasks provided. The last section covered different practical orchestrator use cases and the best practices used during UiPath Orchestrator administration in the ABC Insurance Corporation UiPath CoE.

The core idea of this chapter was to make sure that new UiPath Support team personnel such as Jennifer are able to learn and start undertaking UiPath Orchestrator administration tasks quickly.

I hope this chapter was helpful to get a quick overview of administering UiPath Orchestrator. In the next chapter, we'll discuss further what a typical day for UiPath support personnel looks like and detail different real-life support scenarios.

5

Robot Management and Common Support Activities

This chapter will provide an overview of UiPath Robot administration and help you to understand real-life support activities. Handling UiPath support requests within an enterprise setup is the core responsibility of the UiPath support personnel. A good understanding of request categorization based on priority and complexity is needed for the support team to triage incoming support requests. Support requests are usually managed by an **IT Service Management (ITSM)** system such as ServiceNow or Zoho. Hence, basic training in the ITSM needs to be provided to the UiPath support team to utilize and handle requests in the system.

The sections of the chapters are designed to give the readers context from a practical note. It is done by extending the UiPath support persona introduced in previous chapters. Jennifer from ABC Insurance Corporation's **Robotic Process Automation (RPA)** team will perform different tasks assigned to her in the rest of the chapter and that will be the learning for you.

> **Note**
> In technically mature organizations like ABC Insurance Corporation, the ITSM system ServiceNow is integrated into **Application lifecycle Management (ALM) a**pplications such as Jira to have seamless integration between request, incident, and change management processes.

Here is what you will do as part of this fifth chapter:

- Understand the UiPath Robot management and support details
- Discover the best practices in handling common UiPath support categories of requests
- Learn about real-life UiPath support activities performed by support personnel
- Get to know how a business-impacting service request needs to be handled
- See how to perform periodic support and complex requests that involve external teams

Let's start with robot management detail in the first section of this chapter.

Robot management

Managing UiPath Robots by performing the add, update, and remove Robot tasks is essential for UiPath support personnel. **Unattended Robots** and **Attended UiPath Assistants** are the two UiPath Robot categories discussed. The actual Robot or assistant software is installed in the physical or virtual machines, and the configuration and management of these Robot artifacts are carried out in **UiPath Orchestrator**.

Both the machine and the Robot are configured as tenant-level entities, hence the proper mapping of tenant-level machines and Robots needs to be available in the first place, for example, let's say machine *VD-01 hosts RPAROBOT1 or machine template MT-01 hosts RPAROBOT1, RPAROBOT2, and RPAROBOT3.*

Robot on machines

Before configuring Robots on a machine, we need to ensure that the client UiPath Robot software is installed on the machine. **UiPath Robots or Assistant clients** are connected to the Orchestrator using the Orchestrator URL and the machine key. This information should be available to the support personnel before connecting the Robot machine to the request Orchestrator tenant.

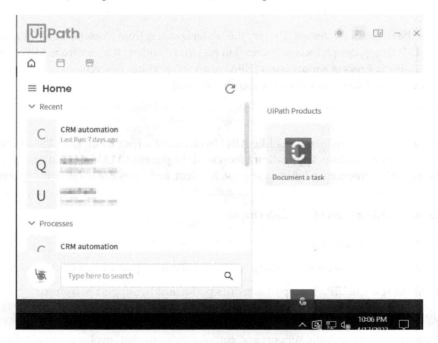

Figure 5.1 – UiPath Assistant

Once the connection is established, then it is the responsibility of the UiPath support team to make sure the Robots in the machines are successfully connected with the green status.

The Robot client's connection can be terminated using the **Disconnect** option in the Orchestrator settings. The support team might need to log into the Robot machines to troubleshoot support issues, and hence it is essential to store the access details of these Robot machines in a centralized and secure location.

Robot details in Orchestrator

A list of all available Robots is accessed from the **Tenant | Robots** tab. This will be the primary place where UiPath support personnel will access information about the Robots during support requests management.

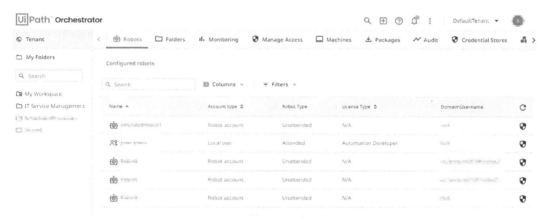

Figure 5.2 – UiPath Robots list

The Robots must be assigned to folders before executing any automation. The Robot needs to be removed at the **Folder** level before being removed at the **Tenant** level. Hence, the UiPath support personnel must understand that Robot management is ideally carried out at the following levels:

- The machine level
- The tenant level
- The folder level

> **Note**
>
> All the Robots and their access details in the ABC Insurance UiPath **Center of Excellence (CoE)** are tracked on a **Confluence** page. The Robot details will have the **Name**, **Robot IDs**, **Tenant**, **Folder** or group mapping, the credentials reference to credential store assets, access to the application, and license details such as Office or Adobe, for example. In matured RPA operations, these details are automatically pulled from the UiPath database and available in a web portal that has options to search and retrieve these details on the fly. That way, the manual maintenance of data can be avoided.

Next, let's learn how support requests are prioritized in the ABC Insurance UiPath support team, before looking at the UiPath Support scenarios.

Support request prioritization

The ABC Insurance Corporation's UiPath team has adopted the industry-standard prioritization model. This model is built on two major criteria:

- Business criticality
- Urgency (from stakeholders)

When a support request or incident is raised, urgency and criticality values will determine the priority of the incident, which dictates the **Service Level Agreement (SLA)**.

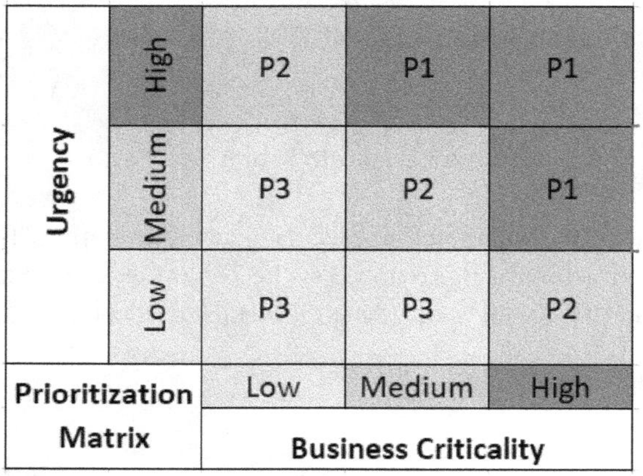

		Low	Medium	High
Urgency	High	P2	P1	P1
	Medium	P3	P2	P1
	Low	P3	P3	P2
Prioritization Matrix		Low	Medium	High
		Business Criticality		

Figure 5.3 – The ABC Insurance RPA support team prioritization matrix

Let's get started with the details of a few P1 support tickets handled by Jennifer in the next section.

P1 – A high-priority support request

A support request with a high impact on business operations and marked as business urgent is marked as a **P1 ticket**. All the support requests flowing through the UiPath support queue will be triaged to confirm their priority level. Priority 1 issues are given the most importance by the UiPath support team and the SLA is shorter than that of other types of requests.

In ABC Insurance Corporation, if a support request is triaged and P1 is confirmed, a dedicated support team member is assigned to that request. All the existing resources will be channeled to this request until the issue is resolved. UiPath developers will also be involved to consult on possible fixes for P1 if required.

> **Note**
> A P1 incident is one of the essential metrics tracked by the UiPath CoE leadership team. SLA adherence and reducing the P1 instances are the two most important objectives for the UiPath support team.

Let's look at a few scenarios of how Jennifer handles production P1 tickets assigned to her.

S1 – All email automation jobs failing on a particular Robot

Process: In ABC Insurance Corporation operations, there is an email box that receives questionnaires related to end user claims. This Claims email inbox is monitored by a dedicated Unattended bot scheduled to run every hour (24/7). The bot will read the incoming email and save the attachment based on the business rule (the email's subject and the sender's address).

Issue: The bot job is getting faulted continuously and a P1 ticket is raised by the UiPath monitoring team based on the job status of UiPath Orchestrator.

Approach and fix: Jennifer disables the trigger of this job and then looks at the Orchestrator logs. She identifies the root cause of this issue, as the bot was failing at the **Get Outlook Mail Messages** step. When she logs into the Robot machine, she finds a pop-up dialog and clears it. The Outlook settings were updated during a system maintenance activity and that introduced a popup requesting the user to **Allow** or **Deny**.

Figure 5.4 – Outlook notification

She cleared this popup and then updated the programmatic access settings in Outlook, for example, **Outlook | Trust Center Settings | Programmatic Access | "Never warn me."** This will ensure that the confirmation dialog will not appear the next time the bot runs.

She enables the trigger and monitors the run until it is completed successfully. Finally, she updates the incident with the resolution and closing comments.

S2 – Business-critical web automation bot down

Process: An Unattended bot was scheduled to perform a few steps on an internal claims web application and pull out the claims that matched a business rule. It was scheduled to run once a day, around noon, and the data was needed for a reconciliation report sent every day. The input file has nearly 5,000 claim numbers, hence it is nearly impossible to be processed manually within a day.

Issue: The first bot run of the week failed with an error message on the Robot machine screen and it was marked as **Faulted** as the exception was not appropriately handled by the script.

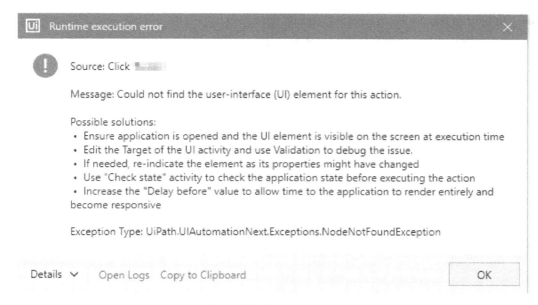

Figure 5.5 – Error message

Approach and fix: Jennifer checks the Orchestrator logs for this process's last run and finds that an unhandled exception related to **Type Into** was failing and throwing the error message. Over the weekend, there has been a change deployed in the target claims management application, and the selector seems to have changed in the text box where the bot enters the claims number.

As this is a change to a script, Jennifer creates an emergency change request to get the selector fixed by the respective UiPath development team. She then updates the comments on the ticket and associates the change request with this P1 ticket. Once the change is implemented and validated, the P1 ticket is closed.

S3 – Attended automation job failed

Process: A claims representative (business user) will trigger an attended bot and pass the claimant details as input parameters. The bot will use the business rules in an Excel file in a shared location to identify the result and respond to the user. This is an on-demand process but is critical to performing the claims adjudication step.

Issue: The bot run failed with an error message on the screen.

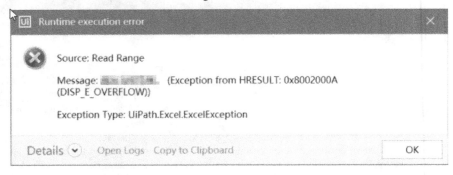

Figure 5.6 – Excel automation exception

Approach and fix: Jennifer starts a remote session on to the business user's machine and she identifies that the issue is related to the **Read Range** activity in Excel. She reviews the Excel file and finds that a few columns read by the bot are missing.

Someone from the business team has accidentally removed the input file fields, which was the root cause of this issue. Jennifer talked to the business stakeholder and corrected the business rules Excel file. She then asked the business user to rerun the job, which was successful.

> **Best practices**
>
> For P1 tickets, you need to ensure the following:
>
> a) You have a dedicated team and to track the SLA.
>
> b) You record the **Root Cause Analysis** (RCA) and solution in the central support documentation.
>
> c) You have dedicated communication channels such as Teams or Slack with relevant stakeholders.

The three P1 incidents discussed in this section might have given a glimpse of real-life P1 UiPath incidents for new support personnel.

In the next section, let's discuss a few medium priorities support requests handled by Jennifer.

P2 – A medium-priority support request

A support request is categorized as a medium priority when the impact on business operation and the urgency to solve it is relatively moderate compared to P1. In the ABC Insurance Corporation RPA support team, medium-priority requests are usually categorized as P2 tickets, and the accepted SLA might be applicable. The UiPath development team can also be consulted for a few P2-type support requests.

Most of the business requests are classified into P2 issues or requests. In ABC Insurance Corporation, if a support request is triaged and P2 is confirmed, anyone from the support team can handle this request. The incident can be assigned to other team members as well.

Let's get into the details of a few P2 support requests handled by Jennifer, which are outlined in the next section.

S1 – A business query on the failed job run

Process: An Unattended bot generates a consolidated **Aged Claims** report based on different business rules from the claims management system. This report is usually uploaded to a cloud share drive. After every run, a notification to the business is sent with the audit report.

Issue: The bot ran on schedule, prepared the report of the aged claim, and uploaded it to the shared drive, but the audit report was not shared with the business stakeholders.

Figure 5.7 – The job status

Approach and fix: Jennifer checks whether the job to be in a **Faulted** state but the output report is already available. Further investigation reveals that the email notification failed due to an Outlook mail activity issue. She starts a remote session on the bot machine and finds the Outlook error popup. She then clears the popup and notes this in her ticket.

As per the business stakeholder's request, she restarts the **Faulted** job manually to generate the audit report, as reprocessing the report will not hurt the business operation.

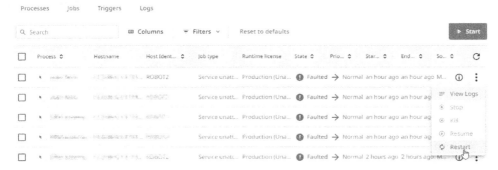

Figure 5.8 – The job context menu

The business operation was not wholly halted because of the bot's issue but because it falls into the P2 category.

> **Note**
> There will be P2 cases where the support team reaches out to the UiPath development team for enactment change requests, such as replacing send mail **Outlook activity** with **Simple Message Transfer Protocol (SMTP)** activity or configurable timeouts.

Let's look at other P2 ticket details next.

S2 – Execution logs are not generated for a machine

Process: An Unattended bot is used to receive wire transaction details from a partner bank and consolidate the data. The data is then fed into a mainframe claims application. It runs every hour during working days.

Issue: This time-critical finance management process (monitored by the UiPath monitoring team) did not create any execution logs in UiPath Orchestrator during or after job execution for a few recent runs.

Figure 5.9 – The logs issue

Approach and fix: Jennifer confirms that the job state is completed or running but that the logs were not generated for the last three runs. She also identified that all the jobs running on a particular machine have the same logs issue. She makes a note of these impacts in her support ticket.

She then refers to the **Support reference guide** link on the **Confluence** page, which contains articles on common issues that other support teams face and appropriate solutions. She spots the relevant article, which provides the following steps to clear the cache and restart the Robot service:

1.   Stop the UiPath Robot service on the machine.
2.   Delete the `C:\~\SysWOW64\config\systemprofile\AppData\Local\UiPath\Logs\execution_log_data` folder.
3.   Restart the UiPath Robot service.

Once the steps were performed, future runs on the machine were producing the execution logs and the ticket was closed.

As this is a business-critical process, losing execution logs will make it impossible to troubleshoot issues and meet SLAs. This proactive monitoring from the team helped to mitigate risk.

S3 – Robot provision or unprovision during ServiceNow maintenance

Process: Five Unattended bots are used to execute ServiceNow-related processes such as creating, updating, or retrieving data from the ServiceNow application.

Issue: There was unplanned downtime for ServiceNow and bot processes were faulting. When the monitoring team reached out to the ServiceNow application team and business owners, they requested to stop all the bot processes related to ServiceNow.

A P2 ticket was created to handle this request, as these were supporting processes.

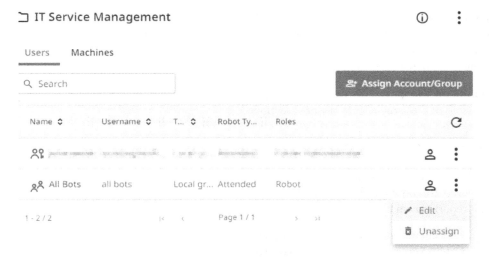

Figure 5.10 – Bot management

Approach and fix: Jennifer found 30 different processes related to the ServiceNow application running in production. Few jobs were scheduled and external systems triggered a few. She kills the running jobs and then disables the related triggers. Still, to stop the jobs triggered by the external system through API calls, she needs to unprovision the Robot users.

She checks and finds that all the processes are listed in an IT Service Management folder and a proper user group called **All Bots** was created. She just unassigns the users of this group and the processes stop.

Once the ServiceNow application is operational, she undoes all the steps to provision the bots and enables the triggers.

> **Best practices**
>
> For P2 tickets, you need to ensure the following:
>
> a) You have business stakeholder and development team details for UiPath processes in a central repository.
>
> b) You claim for reduced support effort by providing business stakeholders with minimum access rights to monitor or rerun failed jobs.
>
> c) You have a central repository for storing UiPath infrastructure details and access details.

Next, let's understand the details of a few P3 support tickets handled by Jennifer.

P3 – A low-priority support request

A support request is categorized as a low priority when the impact on business operation and urgency to solve it is relatively low. In the ABC Insurance Corporation RPA support team, low-priority requests are usually categorized as P3 tickets, and an accepted SLA might also apply. The quantity of P3 requests will usually be high, hence most of the UiPath support personnel's time will be spent on these tickets.

A few common P3 support requests handled by Jennifer are outlined in this chapter section. Let's get into the details of these requests in the next section.

S1 – Managing business users in the audit report

Process: An Unattended bot executes periodic business rule updates in the **customer relationship management (CRM)** system. The changes made by the bot are then shared with the business leadership over audit report via email.

Issue: There is a change in leadership; therefore, a new member needs to be added to the audit report recipient list, and an existing member (John Davis) needs to be removed, as he is no longer in the group.

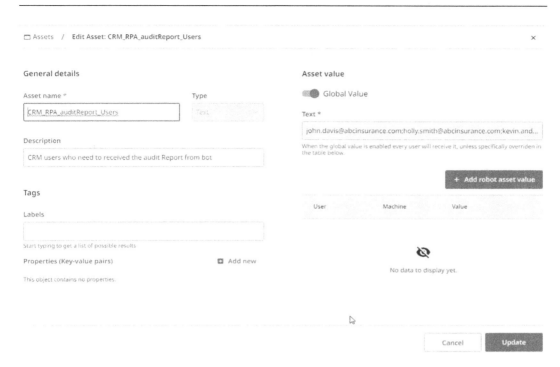

Figure 5.11 – Edit Asset

Approach and fix: Jennifer finds the email IDs of both of the updates from the P3 ticket. She then locates the **Asset name** related to this process from a central UiPath repository mapping file. She edits the same asset in Orchestrator to remove John and add Tim's email to the value.

There is little risk associated with this kind of request, hence they are categorized as P3 tickets.

> **Note**
> Most UiPath support request P3 tickets can be solved without the help of the UiPath developers and it is always advisable to record the standard operating procedures of common support activity requests in a centralized knowledge repository or a tool such as Confluence.

S2 – Advanced trigger management

Process: An Unattended bot to execute the active customer report is currently generated by triggering a bot manually around the first week of the month.

Issue: Due to compliance requirements, the customer report must automatically be extracted at midnight on day one of every month.

> **Note**
>
> A **cron expression** is used to represent a customized schedule for jobs and it is a string that explains the schedule details of the job.

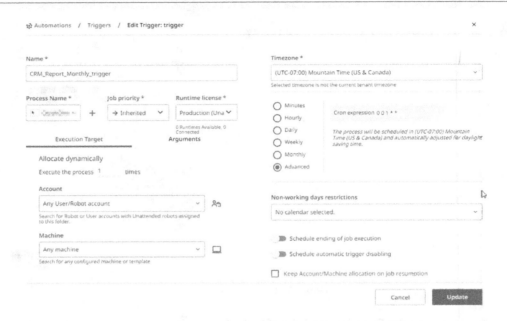

Figure 5.12 – Cron in trigger dialog

Approach and fix: Jennifer adds a new trigger for this monthly scheduled process but the issue is that she cannot use any of the options available within the **Triggers** dialog. She must generate a cron expression and use the advanced option to create this trigger to run on the first of every month at midnight. A cron expression can be generated using online tools. She uses one such utility to generate *0 0 1 * ** and updates the trigger.

> **Note**
>
> Knowledge about cron generators, selectors, credential vaults, data extraction services, SQL, cloud capabilities, and the like is essential for UiPath support personnel.

S3 – Logs check and email alert issues

Process: An Unattended bot is used to update customer details such as name, address, and phone number in the internal CRM system based on an input Excel file that has been placed on a shared drive by the business users. Region-wide customer reports are generated daily and passed on to the sales team. This is accomplished by downstream bot automation, which is triggered based on the completion of this original parent bot process step. There are hundreds of customer details updated every day by this bot process, hence it is a business-critical process step.

Issue: The business user checks and finds that **Account 123456** is missing in the downstream automation and asks UiPath support to look for the root cause of the missing business transaction.

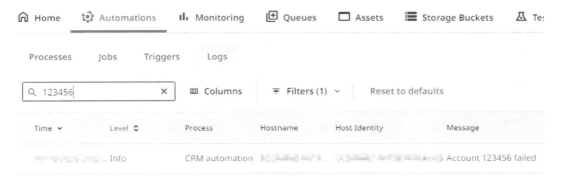

Figure 5.13 – Searching logs

Approach and fix: Jennifer gets the request and quickly searches for the account number in the logs by typing the text in the search box and finds that the bot failed to update the account; the same is also recorded in the audit report. Jennifer notes these details in the support ticket and updates the business user.

Best practices

For P3 tickets, you need to ensure the following:

a) You try to look for patterns in the raised request and create problem requests and change requests to improve existing processes.

b) You aim to reduce support effort by identifying use cases for "self-help" or automated support.

c) You have a customer survey to measure the customer experience of the UiPath stakeholders.

I hope this section was interesting. Now, let us look at the details of a few more support request scenarios.

Other support requests

This section will cover a few ad hoc support requests that fall into complex and periodic UiPath support requests. Let's view the detail in the following section.

Complex support requests

This section of the chapter explains a few complex support requests handled by Jennifer. A support request is categorized as a complex one when external teams are involved in solving an issue. It can be a different team in the same organization or a vendor's support team. Even the UiPath official support team will be involved if there are UiPath platform-specific issues.

Complex requests can be categorized as P1, P2, or P3, and an accepted SLA may also be applicable. These are challenging tasks for the UiPath support personnel and a few scenarios covered in this section will help you prepare to handle these complex UiPath support requests.

> **Note**
>
> As UiPath support personnel, having an overall knowledge of the RPA ecosystem is needed to handle complex support requests.

S1 – Cloud service provider change – Amazon WorkSpaces to Azure Virtual Desktop (VD) for Robot machines

Process: 20 UiPath bot processes executed financial automation under the **Finance** tenant. There are 20 AWS workspaces provisioned as Robot machines with UiPath **Robot/Assistant** client software installed and connected to Orchestrator.

Issue: The finance business unit has recently decided to move from AWS to a Microsoft Azure cloud provider for their business unit infrastructure needs. Hence, all the existing UiPath Robot machines in Amazon WorkSpaces need to be decommissioned, and new Azure VD instances need to be provisioned as Robot machines.

The ABC Insurance cloud infrastructure team has created 20 new Azure VD instances with Robot active directory group permissions and all the software needed for the Robot to operate. The open request now is to configure these new machines and establish a connection with Orchestrator.

Approach and fix: Jennifer receives this request, and as this request needs a downtime of production jobs, she makes sure that the business and IT stakeholders are notified of this bot downtime. The tentative downtime interval and schedule are communicated to the stakeholders.

Jennifer replaces the AWS machines one by one, taking down the AWS machine and spinning up the corresponding Azure machine. This would eliminate downtime and allow her to gracefully back out if it turns out one process or another can't be brought online in the new environment.

She unprovisioned all the Robots in Orchestrator by unassigning the Robot group (in all the folders). She waits until all the running jobs on the Robots are completed and then logs into Azure VD instances and establishes a connection with Orchestrator using the machine key one at a time.

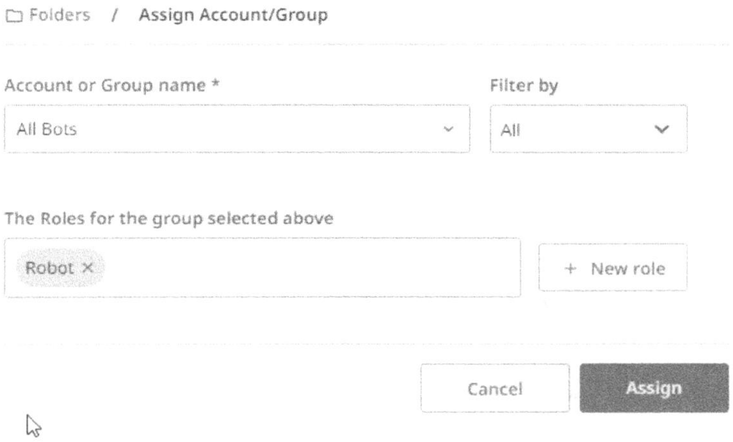

Figure 5.14 – Provision bots

Once all the connections are established, she runs a sanity test job on each machine from Orchestrator. If the sanity test passes, she then provisions one of the folder Robots and monitors the run.

If there are no issues, the Robot is provisioned based on the folders. The runs are monitored closely for a day. The same procedure is followed for the rest of the 19 machines as well before closing the ticket.

S2 – Third-party product integration – Test Manager with ServiceNow

Process: There is an integration between UiPath Test Manager and ServiceNow Test Management to track test automation coverage from ServiceNow. The connection was set up using the API key from the UiPath connection panel.

Issue: The UiPath testing team uses UiPath Test Manager to execute the automated test and make sure all the tests in the ServiceNow system are executed. When they were trying to execute the test, the link to ServiceNow from the **Requirements** tab was soon to be broken, hence they raised a support ticket with Jennifer's team.

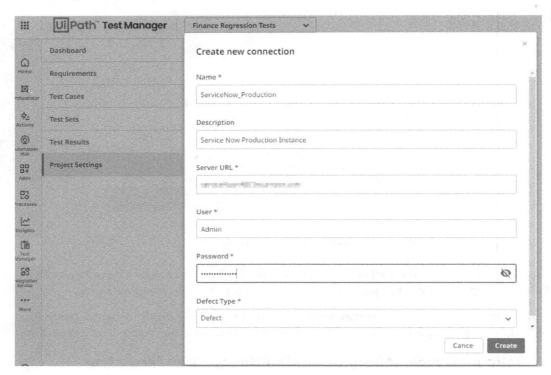

Figure 5.15 – Test Manager connection settings

Approach and fix: Jennifer verifies that the connection is broken and then she sets up a call with stakeholders from the ServiceNow administration team and the UiPath test automation team. With the help of the ServiceNow administration team, they find that the recent change in ServiceNow has removed the **UiPath Test Suite** Integration. Hence, she generates a new API key from UiPath Test Manager and passes it to the ServiceNow administration team, configuring it in ServiceNow. The connection is reestablished and the ticket can be closed after validation.

S3 – UiPath Robot machine performance issue

Process: All of the 20 Robots newly configured in the **Finance** tenant executed the financial RPA processes.

Issue: Robot connections were being dropped intermittently during executions and it is starting to happen in two of the Azure VD machines today.

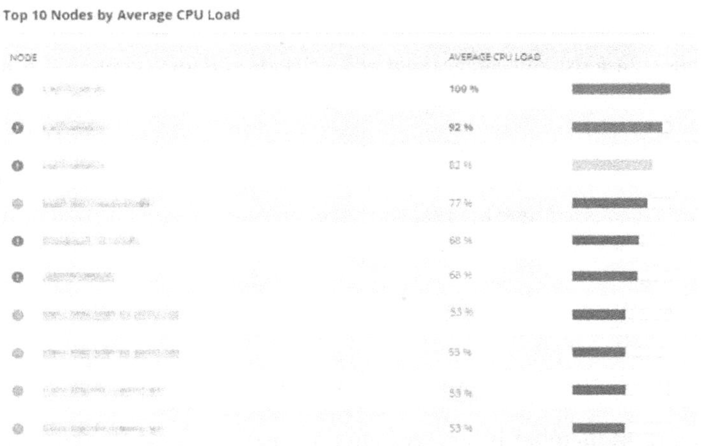

Figure 5.16 – CPU load

Approach and fix: Jennifer checks the jobs and identifies that recently the Robots on those two VD instances were entirely occupied and many jobs were pending in the queue. She quickly unprovisions the Robots and informs the stakeholders. Then, she pulls out the infrastructure monitoring dashboard and checks the CPU load during that period when the connection dropped and validates that the CPU spike has caused this issue on those two VD instances. As a short-term fix, she spreads the load to all the Robots and recommends upgrading the VD instance type with a higher processing configuration.

> **Best practices**
>
> For complex tickets, you need to ensure the following:
>
> a) You have regular catchups with the enterprise architecture team to learn about changes that might affect UiPath infrastructure and bot processes.
>
> b) You build custom monitoring utilities that can help reduce the time it takes to identify the root causes of issues.
>
> c) You make note of all the support tickets related to external teams to identify the areas where we can reduce dependency in the future.

This concludes how Jennifer handled a few common scenarios of complex UiPath support requests. In the last section, let's look at various support requests that may be periodic.

Periodic support requests

A support request is categorized as periodic when a support team's assistance is requested for a few planned activities within the UiPath team. A few common periodic support requests handled by Jennifer are outlined in this section.

Periodic support requests are usually categorized as P3 tickets and an accepted SLA may also apply. Awareness of these possible periodic UiPath support requests will help you prepare for real-life requests.

> **Note**
> Most of the requests discussed here are real work scenarios and having a good understanding of these will provide a headstart to new UiPath support personnel in their work.

Let's get into the few requests in the next section.

S1 – Production validation trial run and hypercare issues

A UiPath monitoring team member is involved after any new process or change is deployed in a production environment to monitor initial trial runs or sanity checks. In many scenarios, additional monitoring is in place for the first few days (called a **Hypercare period**) to validate the new changes.

Recently, Jennifer received a request from a hypercare monitoring team on a business-critical **policy management bot** process change deployed over the weekend. The ticket stated that the SLAs were not met due to intermittent faults in the jobs. She investigates the logs and figures out a particular popup is not handled by the bot (which happens when a particular business rule is met) and has raised a change request.

S2 – Action of application monitoring alerts

All the UiPath processes developed by the ABC Insurance UiPath CoE have exception management. The exception is caught whenever a job faults and an email is trigged to the respective support and business teams.

It is recommended to create separate support tickets for these alerts to investigate the failure's root cause and fix it if the issue is not an intermittent one.

CRM UiPath Failed to Login Inbox ×

to me ▾

Dear Support Team
CRM Automation Failed to login. Please check the error details below:
Exception:Could not find the user-interface (UI) element for this action.

Possible solutions:
• Ensure application is opened and the UI element is visible on the screen at execution time
• Edit the Target of the UI activity and use Validation to debug the issue.
• If needed, re-indicate the element as its properties might have changed
• Use "Check state" activity to check the application state before executing the action
• Increase the "Delay before" value to allow time to the application to render entirely and become responsive
Thanks
Bot2

Figure 5.17 – A monitoring email sample

Jennifer received a support ticket created based on an application monitoring email generated by the CRM bot. She investigates and confirms that the login screen web selectors are changed and creates an high-priority emergency change request to fix the issue.

S3 – User management and access audit

User and role management on the UiPath platform is a common UiPath support request frequently addressed by the UiPath support personnel. In many scenarios, periodic audit checks are performed to view only the access of authorized users to the UiPath resources. All these access-related requests can be handled from UiPath Orchestrator.

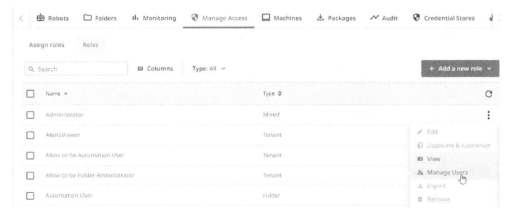

Figure 5.18 – Manage Access tab

There is a request assigned to Jennifer to remove a user from UiPath Orchestrator who is no longer with the UiPath development team. She removes the access to individual folders and tenant levels. Finally, she removes the user from the platform administration user screen too.

S4 – Scalability – Add additional bots to handle the extra volume

Adding additional bots to folders where there are many jobs in a **Pending** state is one periodic request received by the support team. This request gets raised when a volume surge is experienced due to a business demand or a technical issue backlog.

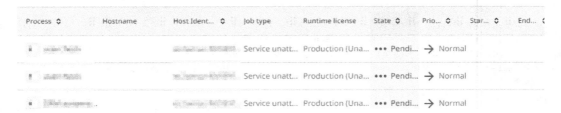

Figure 5.19 – Job status

A few hundred jobs were pending on a particular claims processing process. As there were many queued, an SLA breach was flagged automatically to the support team by the monitoring job. Jennifer was assigned a task to add additional bots that were ideal from a different folder. She removed the bots from the parent folder and assigned them to the group used for this claims processing process. After a few hours, the backlogs were cleared; she moved the bots back to the original parent folder and closed the ticket.

S5 – Data issue troubleshooting

A bot executes numerous jobs, and as most of them deal with production business data, there are circumstances where the stakeholders question the data entered by the bot. Again, this is a periodic request received by the ABC Insurance Corporation UiPath support team.

For instance, in one of Jennifer's requests, the bot entered a phone number in place of a street address in the **Address** field of a mainframe claims application. This issue was raised from data audit findings. She found the root cause when she investigated the bot process logs. There, a particular business rule was met, then the bot navigated to a mainframe screen, and there was an additional field introduced on that screen that made the bot enter the wrong values in the address field. Business stakeholders use a bulk data update to update the wrong information and the bot process was updated with a change request.

Additional periodic requests

The following list is a collection of UiPath support requests that happen as part of the daily routine in the ABC Insurance UiPath support team:

- Onboard or offboard new UiPath developers
- Onboard or offboard new bot process – check prerequisites and provide access
- Triaging, incident diagnosis, and rectification
- Platform maintenance assistance
- Scalability planning and implementation support
- Password expiry of bot accounts to external systems
- Redundancy by having backup bots – replace a Robot with a backup
- Information security incident troubleshooting – questions to check for Robot involvement
- Check why few Robots are not available to execute jobs
- Updates on support systems like upgrades or downtime
- Emergency change deployment support
- Data issues troubleshooting – to check why the Robot inserts invalid data in the application
- Web browser plugin issue – web automation fails due to plugin issues
- Product bot license expiration, such as Adobe or Office

> **Best practices**
>
> For periodic requests, you need to ensure the following:
>
> a) You build support utilities to automate common periodic support activities wherever possible.
>
> b) Having a shared support team calendar with important periodic support activities will be helpful.
>
> c) Having a central repository with all UiPath product and support application license information with expiration and renewal details will help fast-track support activities.

These are some of the real-life UiPath support requests raised periodically and these kind of requests will becomes part of the everyday work routine for UiPath support personnel.

We have walked through and understood how Jennifer handled the UiPath support tasks assigned to her. It's time to recap what we learned in the following section.

Summary

Having a practical understanding of how to manage Robots and real-life use cases in UiPath support operations is vital for any new UiPath support personnel. We started this chapter with an introduction to Robot management. The first section covered all the basics of how a Robot entity in Orchestrator is related to the software client on a virtual machine.

The support prioritization framework used by ABC Insurance Corporation was introduced before we got into the different support ticket details. The following section discussed the core functions of handling P1 high-priority support tickets with three use cases. The same pattern was followed in the next two sections, which discussed medium-priority and low-priority support requests.

In the next section, a few complex support requests were discussed to outline the challenging aspects of this UiPath support role. Finally, many periodic support requests were outlined to give you different flavors of the support requests you might encounter.

Although this chapter covered only a handful of support scenarios concerning ABC Insurance Corporation, what we covered will give new support personnel some insight and direction when navigating the role's responsibilities.

Best practices were outlined in each section to add more context to performing these tasks in the best possible way. The core idea of this chapter was to make sure new UiPath support team personnel understand how typical UiPath support personnel would work requests on a large-scale UiPath operation.

I hope this chapter helped give a quick overview of performing Robot management and handling UiPath support requests. Let's discuss the very interesting concept of DevOps in UiPath in the next chapter.

6
DevOps in UiPath

There are different software development methodologies, and DevOps is one of the recent trends that makes sure quality products are delivered using rapid and continuous development through integration, testing, deployment, and monitoring phases.

This chapter will provide an overview of UiPath deployments and introduce DevOps concepts in UiPath programs. In addition to providing an overview of the overall automated delivery process, this chapter will also introduce the reader to all five phases of the DevOps lifecycle using various application tools, such as GitHub, Jenkins, the UiPath Jenkins plugin, and Insights. Here are the main topics of the process:

- Continuous Development

- Continuous Integration

- Continuous Testing

- Continuous Delivery

- Continuous Monitoring

Finally, the chapter ends with how DevOps principles are used in the ABC Insurance UiPath CoE.

At ABC Insurance, the UiPath support team also helps to troubleshoot deployment issues. Hence, all the members must go through the DevOps training and understand the existing setup. We will explain to Jennifer how the UiPath release management in the ABC Insurance Corporation started their journey, introduce the DevOps phases, and how the program ultimately became mature in adopting the DevOps phases.

> **Note**
> In technically mature organizations such as ABC Insurance, the ITSM system ServiceNow is integrated with DevOps applications such as Jenkins or Circle CI to have seamless integration between change management and release management processes.

Here is what you will learn as part of this sixth chapter:

- You will learn about the UiPath deployment process and best practices.
- You will understand the DevOps lifecycle and the tools concerning the UiPath program.
- You will learn how GitHub is used in Continuous Development and Integration.
- You will understand how Jenkins is used in continuous testing and deployment.
- You will understand how continuous monitoring works.
- You will get to know how the ABC Insurance Corporation uses DevOps in the UiPath program.

Let's understand the UiPath deployment process and its details in the first section of this chapter.

UiPath build and deployment management

The UiPath scripts developed by the UiPath developers will be published as NuGet packages in UiPath Orchestrator. This initial step in creating the NuGet package is the build step. The NuGet packages are the basis for creating the automation process in Orchestrator. The work steps involved in making a UiPath NuGet package deployed as an UiPath process that can be executed as the job (by robots) is referred to here as the UiPath deployment process.

When the UiPath program was at the initial level of maturity in the ABC Insurance UiPath CoE, there was just one development environment. The UiPath scripts were developed in this environment and a production environment was tied to Orchestrator where NuGet packages were published to the Orchestrator tenant.

There were no proper script reviews and external testing teams involved; the bots had many quality issues and had to be fixed constantly. Hence, two more environments using test orchestrators and UAT orchestrators were set up. These changes improved the quality of the bots in the long run and relatively, there were fewer bot production issues than in the previous state.

In the next section, let's discuss how the manual deployment process was operated in the ABC Insurance UiPath CoE team.

The ABC Insurance UiPath manual build and deployment process

The UiPath build and deployment process is triggered when a peer UiPath developer reviews the script developed by the developer and approval is given to create the test NuGet package. The UiPath developer then publishes the package to the test orchestrator. The UiPath administrator, who is part of the ABC Insurance UiPath support team, will create the required resources, such as processes, assets, triggers, queues, and the like, in the test orchestrator and assign a run environment for the bot process.

The bot testing team will use this process to perform their quality assurance activity. If any bugs are identified, the package is rejected, and the process is terminated. Once the testing approval is provided, the same package and setup need to be replicated in the UAT environment, and the UAT team is notified to test them. If the UAT is rejected, then again, the deployment process is terminated. Otherwise, the package is promoted for the production environment.

When the production deployment window is open, the bot administrator will deploy the package to the production orchestrator, and the relative resource for the process is updated. Once these steps are done, the business user or the product owner is requested to validate the production runs. The bot deployment process is completed if the runs are validated and updated in the change management system.

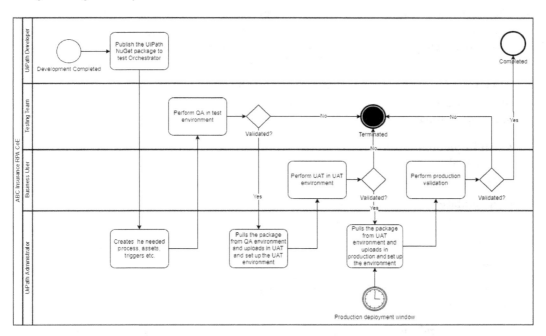

Figure 6.1 – The ABC Insurance UiPath CoE manual build and deployment process

> **Note**
>
> In the ABC Insurance UiPath CoE, the UiPath NuGet packages are stored in a centralized artifact or binary repository manager called **Jfrog Artifactory**. This way, the packages are version controlled, and this is good for auditing purposes as well or for rollback scenarios too.

A few documents are required to perform the manual deployment process. Let's look at them in the next section.

Artifacts used in manual deployments

The artifacts used in the manual deployment process are published in a central repository and should be accessible to all the relevant stakeholders. UiPath Support and the administrator team are usually involved in validating the content of the artifacts. The list of artifacts used in the ABC Insurance UiPath deployments are listed here:

- The **deployment calendar** – A list of all the planned deployments on a release schedule with additional details, such as change requests, stakeholders, status, and so on, are available in this calendar. This artifact will be used to plan the resources for the planned changes and its subsequent business validation.

- The **deployment checklist** – All the deployed processes have to be clear on all the items listed in the deployment checklist. This checklist is usually tied to the change request to check for all mandatory prerequisites for deployment, such as UAT signoff, package availability in a central repository, documents in the release plan, business validator availability, and the like.

- The **release plans** – A set of documents that aids to plan for a UiPath release is termed a release plan. There are three artifacts used by the ABC Insurance UiPath team during release planning, and they are listed here:

  - An **implementation plan** – Step-by-step instructions on how to deploy the package, which also contain information that helps configure the relevant resources in the production environment.

  - A **validation plan** – This artifact lists this artifact's instructions to perform the business validation and relevant scenarios to sign off the package.

  - A **rollback plan** – Step-by-step instructions on how to roll back the package and the configuration in the production environment.

- The **release notes** – A list of all the changes planned to be deployed in a release window is documented in the release notes and it is usually shared before the actual deployment is scheduled.

- **Deployment and validation summary** – This is a summary document usually created to list all the deployment and validation status of the planned UiPath deployments. It is necessary to archive these documents for reporting and auditing requests.

Next, let's learn more about how the UiPath support team is involved in this deployment process.

The support team's role in a manual deployment process

The ABC Insurance Corporation's UiPath team is a group of UiPath administrators involved in the deployment process. Apart from performing the major deployment tasks discussed in the preceding process, they are also involved in a few additional tasks listed here:

- Troubleshooting issues or performing package rollbacks during deployment tasks, involving the development team if needed

- Verifying all the prerequisite accesses and artifacts are available as per the deployment checklist

- Checking whether the change request tied to the UiPath deployment is approved before the actual deployment process

- Communicating release notes and validation status that cover the successful/failed or postponed statuses of deployments to relevant stakeholders

- Updating the central repository of the UiPath bot processes with the latest deployment process information

- Supporting the initial production pilot runs, business validation efforts from business stakeholders, and handover monitoring efforts

---

Best practices

There must be a close association between change management and deployment management in a ITSM tool such as ServiceNow. Make sure to create a **Deployment task** in the production change request of an automation project assigned to the right UiPath administration team.

Now, Jennifer understands the history of the manual deployment process. In the next section, we will introduce the main concept of DevOps and its life cycle phases.

DevOps and its lifecycle

Software Development (Dev) + operations (Ops)= DevOps

DevOps is a Continuous Delivery methodology that enables siloed teams such as software development, testing, release, operations, and the like to coordinate and collaborate to produce quality software products in short-period iterations. The concept has gained traction in every field of software, including the UiPath RPA programs. Let's understand the significance of this DevOps lifecycle before getting into the details.

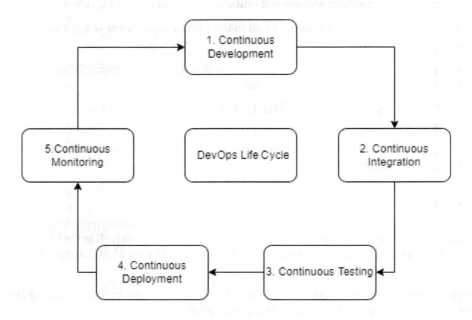

Figure 6.2 – The DevOps lifecycle

Making the process of how software is developed, integrated, tested, deployed, and monitored visible and transparent will be beneficial for the software development team and its other partner teams. This action will ensure that resources and budgets are aligned to support the product.

Having centralized governance for software product development and release process, in addition to shorter release cycles, is another major benefit of this DevOps lifecycle.

Jenkins is the DevOps application used in the ABC Insurance UiPath CoE. Jenkins is an open source DevOps application that helps automate the build, test, and deployment of software productions by facilitating Continuous Integration, continuous testing, and Continuous Delivery of the DevOps lifecycle.

> **Note**
>
> DevOps is one of the fastest-growing segments in the software market, and it is a generic concept that's applicable to any software domain. A wide range of concepts and tools are available for the UiPath development and operations teams. Hence, it is important to choose the right tools that fit the UiPath program goals and are in line with the overall enterprise architecture of the company.

Next, let's start to get into the details of each of these five phases in the following sections.

Continuous Development and Continuous Integration

The first two phases of the DevOps application are the core features that deal with the persona of UiPath developers. The UiPath administrator must understand these phases, as they will be involved in supporting issues that deal with committing the UiPath script or cloning an existing GitHub repository and such.

These concepts will be discussed in detail in the next sections; we will start with some details of how Continuous Development concepts are used in the UiPath automation development process.

Continuous Development of UiPath automation

Continuous Development is an agile software development methodology that will enable an automation program to be split into manageable components. Automation increments are delivered in the form of scripts at different phases of the development process, like so:

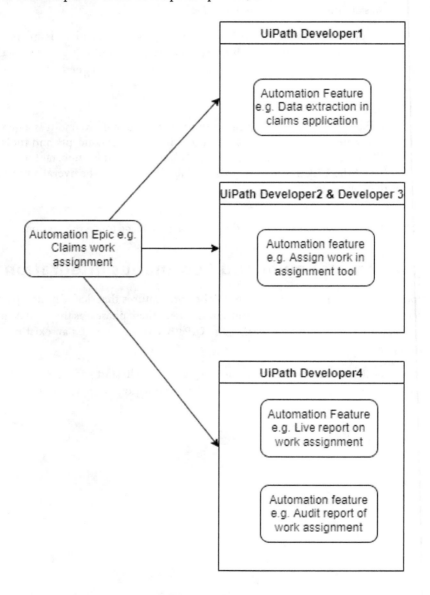

Figure 6.3 – An epic/feature distribution for development

Agile software development is the most popular Continuous Development approach, and it is also followed by the ABC Insurance UiPath CoE as its primary software development methodology. As the UiPath automation deliverables are broken into different components, this type of delivery in the UiPath programs will accelerate the time to complete the development and commit value-driven components. This practice will also encourage using best scripting practices and build a culture of reusability of automation scripts.

Figure 6.4 – Version control options

The version control of UiPath scripts can be performed out of the box using the UiPath Studio's team capability. There are three types of version control platforms supported by the UiPath Studio:

- **Git**
- **Microsoft Team Foundation Server (TFS)**
- **Apache Subversion (SVN)**

> **Best practices**
>
> It is good to have a dedicated central repository for storing these UiPath scripts. In the ABC Insurance UiPath CoE, the UiPath scripts developed by different developers are stored in a distributed version control and source code management platform called GitHub.

The UiPath administrator usually gets involved in the setup of various UiPath Studio development environments, with the GitHub credentials for the developer, the access control for UiPath script repositories in GitHub, and the like.

We will use the **Claims** work assignment UiPath project as an example to explain the rest of the DevOps lifecycle and understand how the scripts developed by different UiPath developers are merged, built, tested, and deployed.

> **Note**
> UiPath Studio has an inbuilt static code analyzer, which the developers need to use to make sure consistent scripting practices are followed across developers. Please refer to this link for more details: `https://docs.uipath.com/studio/docs/about-workflow-analyzer`.

In the next section, let's discuss how version control and integration are performed using GitHub in the ABC Insurance Corporation.

Continuous Integration with GitHub

Continuous Integration is the practice of integrating different pieces of code from multiple developers seamlessly and building a software package that can be tested and deployed. The same principle applies to UiPath script development performed in a distributed setup.

In the ABC Insurance UiPath CoE, the GitHub repository of each UiPath process is created by the concerned GitHub administrator. The UiPath support administrator will also have the same details.

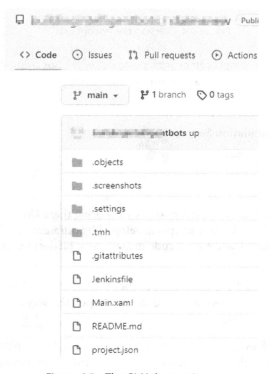

Figure 6.5 – The GitHub repository

There are different branches created by the GitHub administrator, such as development, test, UAT, and production to support the scripted movement between the respective UiPath environments.

Typically, a developer will work on a particular feature (for example, the audit report for a work assignment) of the UiPath automation. The script needs to be committed to a GitHub called a Feature branch. There can be multiple feature branches that make up a primary development branch if multiple UiPath developers are working on different automation features. This concept is referred to as the **branching strategy** in DevOps.

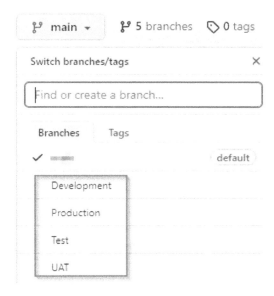

Figure 6.6 – GitHub branches

Once the UiPath script GitHub repository details are provided to the concerned UiPath developers working on the project, they will start by cloning the GitHub repository and then going ahead with the UiPath script development. UiPath Studio has options to commit, pull, and push scripts to and from the GitHub repository using the UiPath Studio environment. It is recommended to provide proper commit comments during each script commit for audit purposes.

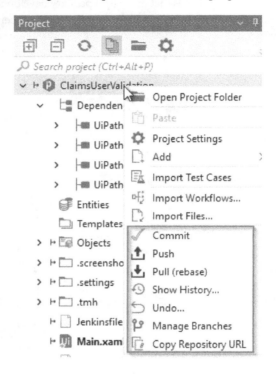

Figure 6.7 – GitHub options

The UiPath administrator will typically be the liaison between the UiPath developer and the DevOps GitHub administrator to help with the right branching strategy for the UiPath program. They are also involved in resolving branch merge conflicts, hence it is good to know a few details about this topic to perform these support and administration tasks.

> **Note**
>
> UiPath script merges in various GitHub branches are controlled by branch rules that can be set up on each branch in the repository. It is a good way to restrict unauthorized script merges by developers and encourage script peer reviews and merges to be made only by authorized users. If there are conflicts during the merges, then it is necessary to have all the developers on the same page and make sure the conflicts are resolved during merge operations.

If we extend the Claims work assignment use case, all the scripts developed by the four developers are committed to four different feature branches. After peer review, they are merged into a development branch. Once the scripts are in place, the most important capability of Continuous Integration is to build a UiPath NuGet package using the scripts available on GitHub. Let's discuss the relevant details in the next section.

Automated build using the UiPath Jenkins plugin

UiPath uses a Jenkins plugin for support, and it can be installed in the Jenkins environment from the plugin's manager. Once the plugin is installed, Jenkins can support the UiPath DevOps function to build and deploy automation.

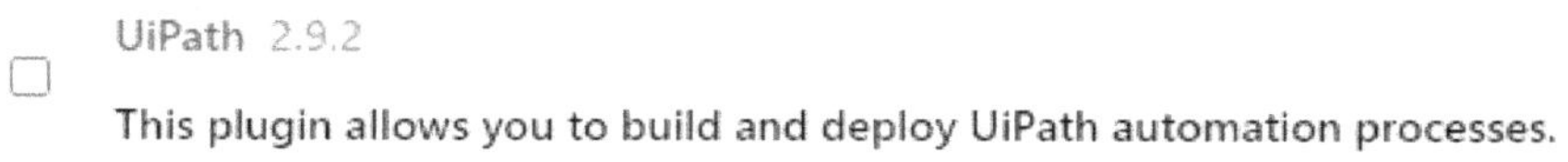

UiPath 2.9.2

This plugin allows you to build and deploy UiPath automation processes.

Figure 6.8 – The Jenkins UiPath plugin

Once the UiPath script is committed to the GitHub branches, then we can try to use the UiPath Jenkins plugin **UiPath Pack** feature to generate the NuGet package of the target UiPath automation.

In the ABC Insurance Corporation UiPath CoE, a **sandbox environment** is set up with Jenkins, and the UiPath plugin is already installed in this setup. This environment will be a good place for support team members to understand how the plugin feature works. Create a **Freestyle** Jenkins project and include a source code GitHub URL. The GitHub repository credentials are already preconfigured in Jenkins.

> **Note**
> Jenkins needs to access the GitHub repository to access the content of the folders. The GitHub repository credentials need to be preconfigured in Jenkins before we can start performing the build operations.

The next important section is the build section, where we can specify the versioning method, the project path, and the output path where the NuGet package should be placed. The output type of projects can be any of these:

- Process
- Library
- Test
- Object project

We will select the process project type to understand how the build process works.

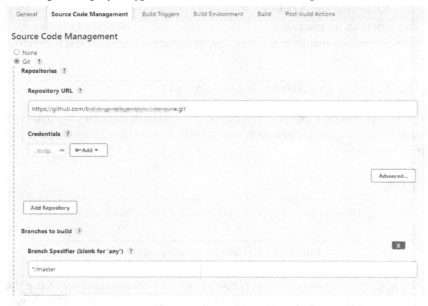

Figure 6.9 – Jenkins source code options

Choose the **UiPath Pack** feature in the Freestyle project's build section and provide the input and output parameters needed to build a UiPath NuGet package.

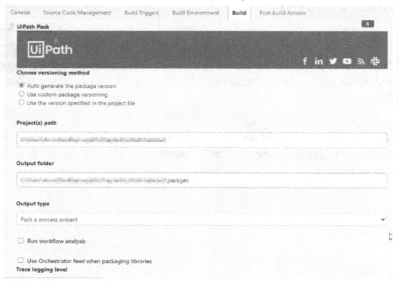

Figure 6.10 – Jenkins UiPath Pack options

Now, save the Freestyle project and click **Build Now** to check if the build process works. It should complete successfully with a NuGet package generated in the output folder.

> **Note**
>
> Besides **UiPath Pack**, there is a UiPath **Manage Assets** feature enabled in the Jenkins Build phase, which can be used to manage the assets of UiPath Orchestrator.

The same principle is converted into the Jenkins pipeline to be generalized and consumed at the UiPath CoE level.

For an enterprise-grade DevOps setup, it is recommended to have a **Pipeline** Jenkins item, and this is an additional step in specifying the build configuration. Working with the DevOps administrator before a Jenkins configuration file is created is recommended. Jenkins configurations are specified in this Jenkinsfile. The file needs to be placed in the project folder and committed along with the UiPath script to the respective GitHub repository. As support personnel will troubleshoot issues with the build, it is good that they know the overall flow of this configuration.

Figure 6.11 – The Jenkinsfile

To get the information for the Jenkinsfile, please browse the **Tenants** page of UiPath Orchestrator. The Orchestrator URL, logical name, tenant name, and folder name details are gathered from this screen and used in updating the Jenkinsfile.

Figure 6.12 – The tenant API key and values

Sample Jenkins pipeline for the build process in Jenkinsfile

Now that we have all the details needed for building a UiPath project using a Jenkinsfile, let's take a look at a sample here:

```
pipeline {
        agent any
        environment {
            ORCHESTRATOR_URL = https://cloud.uipath.com/~/
orchestrator_/"
            ORCHESTRATOR_LOGICAL_NAME = "XXXX"
            ORCHESTRATOR_TENANT_NAME = "YYYY"
            ORCHESTRATOR_FOLDER_NAME = "ZZZZ"
        }
stages {
```

```
    // Building the package
      stage('Build') {
          steps {
          echo "Building..with ${WORKSPACE}"
              UiPathPack (
                  traceLevel: 'None',
                outputPath: "Output\\${env.BUILD_
NUMBER}",

                  projectJsonPath: "project.json",
                  version: CurrentVersion(),
                  useOrchestrator: false
                    )
                }
            }
    }
```

If the pipeline Jenkins project is triggered for a GitHub repository with a Jenkinsfile, then the build steps of the UiPath process will automatically be triggered.

In the casework assignment use case, the NuGet package was generated using this automated build process and stored in the Jfrog Artifactory.

Having a good understanding of the Jenkins pipeline content will save time for troubleshooting Jenkins build issues in UiPath Support and enable faster UiPath development cycles, better visibility of the current state of the project, and faster notification when broken UiPath builds need to be fixed.

> **Best practices**
>
> Integrating the automated build trigger with each merge to the GitHub branch will save a lot of time compared to triggering the build process manually. It will put us in the right direction toward a completely automated UiPath build and deployment process.

In the next section, let's discuss the last three phases of the DevOps lifecycle and how a UiPath CoE can leverage them to mature its technical capability index.

Continuous testing, deployment, and monitoring

The last three phases of the DevOps lifecycle are the core features that deal with UiPath testing, release, and monitoring teams. The UiPath administrator must understand these phases. They will be involved in supporting issues such as when a UiPath package cannot be deployed to a test environment, the automated testing job didn't execute as intended, and the like.

Ideally, once the script is pulled into a GitHub test branch, an automated testing job will be triggered by the Jenkins pipeline. Once the testing jobs are completed, the package is built and deployed to the Orchestrator **Test** environment for additional testing. A similar kind of automated test and deployment pipeline is also configured for UAT and production UiPath environments.

Let's understand how the UiPath Jenkins plugin tests and deploys automation in the next sections to complement the DevOps phases.

Continuous testing with UiPath Test Suite

Continuous testing consists of automated sub-test components applied at various phases of the bot development process, such as unit testing, functional testing, integration testing, and regression testing.

Having automated test automation is a valuable DevOps capability that can be leveraged in this phase. In the ABC Insurance UiPath CoE, UiPath Test Suite generates the bot test automation processes. Let's try to understand the high-level capabilities of Test Suite.

UiPath Test Suite

Test Suite is an automated test automation product offering from the UiPath suite. Test Suite is used to create the test automation processes that can be executed as part of the continuous testing phase of DevOps.

Once the development Studio Pro environment is enabled with test suite capability, UiPath developers will be able to create test automation on the existing UiPath projects.

Figure 6.13 – UiPath Test Suite

The automated tests are created in UiPath Studio Pro and then published to UiPath Orchestrator. The test process is created using the uploaded automated test package. The automated tests are tied to test cases grouped as test sets. The test sets can be executed on a UiPath Test Robot.

Let's extend the Claims work assignment automation case and assume that there are three different test cases, like so:

- An assignment UI login check

- A report generation test

- An email notification test automated as part of this project

These three test cases can be published into Orchestrator and grouped into the Claims test set. They can be executed as part of automated regression tests during any future changes to the bot processes.

Automated testing using the UiPath Jenkins plugin

Like in the previous phase, the UiPath support team member can understand how to trigger an automated test as part of the Jenkins plugin by extending the Freestyle Jenkins project. The UiPath tests are available in UiPath Orchestrator. The Jenkins credential item with the UiPath Orchestrator API key needs to be configured to access these test sets in Orchestrator.

Choosing the right UiPath Orchestrator that points to the right environment is an important part of running the tests. For instance, if UiPath tests need to be executed on the **Test** environment, the relevant UiPath Orchestrator URL must be passed as the input parameter.

Figure 6.14 – Jenkins credentials for UiPath

Once the credential item is updated, the next step is to include the **Post-build Actions** and choose **UiPath Run Tests**.

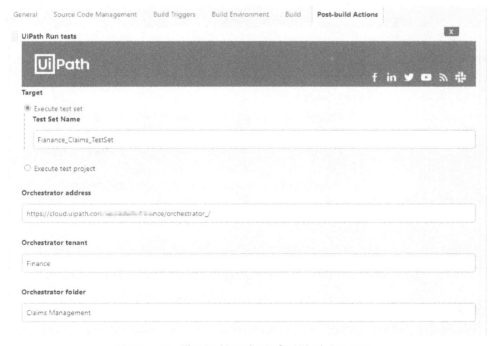

Figure 6.15 – The Jenkins plugin for UiPath Run tests

Once all the input and output parameters are provided, the build now button once clicked, will trigger the UiPath test in Orchestrator.

When we flip to the more practical aspect of the enterprise-grade DevOps setup, it is recommended to have a Pipeline Jenkins item. Jenkins configurations are specified in this Jenkinsfile, which needs to be updated with the **Run tests** information. This file needs to be placed in the project folder and committed along with the UiPath script to the respective GitHub repository.

Once the details of the UiPath tests are available in the Jenkinsfile, the tests can be automatically triggered based on any GitHub action, such as a pull or a merge

Sample Jenkin Pipeline for the Run Test process in Jenkinsfile

Let's look at a sample Jenkinsfile with test steps here:

```
pipeline {
        agent any
        environment {
            ORCHESTRATOR_URL = "https://cloud.uipath.com/~/
orchestrator_/"
            ORCHESTRATOR_LOGICAL_NAME = "XXXX"
            ORCHESTRATOR_TENANT_NAME = "YYYY"
            ORCHESTRATOR_FOLDER_NAME = "ZZZZ"
        }
stages {
            stage('Test') {
                UiPathTest (
                    testTarget: [$class: 'TestSetEntry',
testSet: "TTTT"],
                    orchestratorAddress: "${ORCHESTRATOR_
URL}",
                    orchestratorTenant: "${ORCHESTRATOR_
TENANT_NAME}",
                    folderName: "${ORCHESTRATOR_FOLDER_
NAME}",
                    timeout: 10000,
                    testResultsOutputPath: "result.xml",
                    credentials: Token(accountName:
"${ORCHESTRATOR_LOGICAL_NAME}", credentialsId: 'KKK'),
                                                )
```

```
                              }

    }
```

> **Best practices**
>
> UiPath tests can be triggered per the DevOps pipeline as decided by the UiPath leadership team. In the ABC Insurance UiPath CoE, the UiPath automated tests are executed once the test environment's build and deployment process is complete. Once the automated tests are passed, the UAT deployment process will be triggered.

Improving bot automation quality, reducing bot testing time, and freeing up manual testing hours are the major benefits of this approach for the UiPath automation program.

In the ABC Insurance UiPath CoE, these different types of automated test cases are executed in UiPath environments:

- Functional automated tests (executed in the test environment)
- UAT tests (executed in the UAT environment)
- Regression tests (executed in production environments)

We will cover the most interesting Continuous Deployment concept in the next section.

Continuous Deployment with the UiPath Jenkins plugin

Continuous Deployment is an automated approach to deploying software packages to different environments, such as test, UAT, and production in a pipeline fashion, with audit and configuration management capability enabled at each step of this process.

Multiple UiPath processes can be continuously deployed to various UiPath environments using this approach. Deploying the UiPath NuGet package automatically will help the UiPath CoE mature its ways of working. It will enable the UiPath support team to scale up UiPath CoE deployments without adding additional resources.

In the next section, let's discuss how Jenkins is used in the ABC Insurance UiPath CoE for automated deployments.

Automated deployment using the UiPath Jenkins plugin

In the ABC Insurance UiPath CoE, once an UiPath project is built and the NuGet package is available in a centralized repository, the package needs to be deployed in the test environment.

As in the previous two phases, the UiPath support team member can understand how to trigger an automated deployment as part of the Jenkins plugin by extending the Freestyle Jenkins project. The packages are deployed into UiPath Orchestrator, hence Jenkins plugin needs access to UiPath Orchestrator. The UiPath Orchestrator API is preconfigured with the access using the Jenkins credentials for this reason.

UiPath Deploy can be added in the **Post-build Actions** section of the Jenkins Freestyle project. Test, UAT, and production will have individual UiPath Orchestrator configurations in a typical setup. Jenkins needs to point to these UiPath Orchestrator URLs to deploy the NuGet packages.

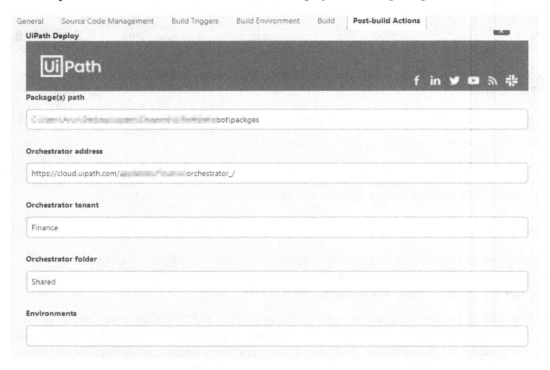

Figure 6.16 – The Jenkins plugin for UiPath Deploy

Once all the input and output parameters are provided, the **build now** button click will trigger the **UiPath Deploy** phase in Orchestrator. Ideally, the package needs to be deployed, and the existing version of the process will also be updated to point to this latest package.

> **Note**
> If UiPath Orchestrator is still using a **Classic** folder, the **Environments** details need to be specified. In a modern folder environment, the **Environments** value can be left blank.

In the actual enterprise-grade DevOps setup where a Pipeline Jenkins project type is at play, Jenkins configurations are specified in this Jenkinsfile that need to be updated with the **UiPath Deploy** information. This file needs to be placed in the project folder and committed along with the UiPath script to the respective GitHub repository.

Once the details of UiPath deployment are available in the Jenkinsfile, deployment can be automatically triggered based on any action on GitHub , such as a pull or a merge.

Sample Jenkin Pipeline for the deployment process in Jenkinsfile

Let's look at a sample Jenkins pipeline for deployment here:

```
pipeline {
        agent any
        environment {
            ORCHESTRATOR_URL = "https://cloud.uipath.com/~/
orchestrator_/"
            ORCHESTRATOR_LOGICAL_NAME = "XXXX"
            ORCHESTRATOR_TENANT_NAME = "YYYY"
            ORCHESTRATOR_FOLDER_NAME = "ZZZZ"
                    }
stages {
stage('Deploy') {
                UiPathDeploy (
                    packagePath: "Output\\${env.BUILD_NUMBER}",
                    orchestratorAddress: "${ORCHESTRATOR_URL}",
                    orchestratorTenant: "${ORCHESTRATOR_TENANT_
NAME}",
                    folderName: "${ORCHESTRATOR_FOLDER_NAME}",
                     environments: '',
                    credentials: Token(accountName:
"${ORCHESTRATOR_LOGICAL_NAME}", credentialsId: 'KKK'),
                traceLevel: 'None',
                entryPointPaths: 'Main.xaml'
                    }
        }
```

> **Note**
>
> Apart from **UiPath Deploy** and **UiPath Run tests**, there is a third option called the **UiPath Run Jobs** feature enabled in the Jenkins post-build phase, which can be used to run a particular UiPath job in UiPath Orchestrator.

Continuous deployments benefit the UiPath CoE by responding to immediate business needs with quick turnarounds on the UiPath automation package deployments in different environments and reducing the risk of wrong deployments happening in production environments.

It is recommended that UiPath support personnel are aware of the continuous deployment process and tools. They might be involved in troubleshooting issues along with the DevOps teams if there are any issues during the UiPath release process.

In the case of the Claims work assignment, the NuGet package is downloaded from the central artifact repository, such as Jfrog Artifactory, and deployed to the respective environment listed in the configuration file.

> **Best practices**
>
> It is a best practice to sync the automated deployment to the production environment to the deployment window listed in the change request, and this can be accomplished by integrating the ServiceNow to the DevOps pipeline, for example, Jenkins.

Let's discuss continuous monitoring concepts in the next section.

Continuous monitoring

Continuous monitoring is used to measure and report on bot application operational performance and the infrastructure in which the applications are hosted. They are used to mitigate the risk of unplanned application and infrastructure downtime and for continuous improvement initiatives.

Various frameworks are followed in different organizations in setting up continuous monitoring in the UiPath RPA program. Let's look at the framework used in the ABC Insurance UiPath CoE in this section:

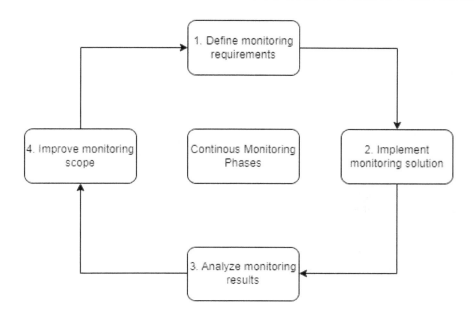

Figure 6.17 – The ABC Insurance UiPath CoE continuous monitoring phases

This framework follows these steps:

1. **Define monitoring requirements**: Monitoring requirements are defined along with the bot development requirements when the bot is being developed for the first time or changed. There might be default requirements, such as integrating the bot process to an existing monitoring dashboard setup with default metrics, such as average processing time, cost saved, and so on. It could also be a custom requirement, such as capturing some process-specific metrics, such as documents processed, user(s) added, and so on. This requirements elicitation process is ongoing even after the bot monitoring solution is placed in production.

2. **Implement monitoring solution**: Once the monitoring requirements are finalized, the proposed solution is implemented, and an improvement to the solutions is identified, they are made in iterations throughout the bot's lifecycle. Five metrics were measured in the initial version of the bot process dashboard; the elegant solution might consolidate the results of the five metrics into two key metrics that make sense to the business stakeholder.

3. **Analyze monitoring results**: The next phase deals with analyzing the monitoring results on an automated continuous loop, which will be used to report on risk items even before they break an existing bot solution. Continuous reporting on a bot server CPU or disk usage are examples that will make sense in this phase.

4. **Improve monitoring scope**: This is the most important phase, where insights from the previous phase are fed, planned, and improvements are conceptualized. For instance, if a particular bot's CPU usage is high on a particular day in a week, which leads to slower processing time, then it is better to have a better scheduling logic to split the load of the bot machine to improve the processing time of the process.

Having this continuous loop of monitoring throughout the UiPath application and platform will improve the operation performance by increasing bot and infrastructure availability, reducing cost by minimizing human error during monitoring and reporting, and increasing visibility, which boosts confidence in the reliability of the UiPath automation program.

If we extend the Claims work assignment, process-specific monitoring dashboards are defined after the process is deployed in the production environment. This is the last step before completing the DevOps lifecycle for this UiPath process.

We will cover details on each of the monitoring phases and how UiPath Insights and Splunk are leveraged in the ABC Insurance UiPath CoE next in *Chapter 7*.

> **Best practices**
>
> It is important to have a centralized dashboard and reporting strategy for the UiPath bot and its infrastructure, using applications such as UiPath Insights, Kibana, or Power BI. Otherwise, siloed monitoring solutions will be hard to support and maintain in the long run.

This concludes all the DevOps training and information for Jennifer. Next, let's see how the DevOps phases are practically used in the ABC Insurance UiPath CoE.

DevOps in action in the ABC Insurance UiPath CoE

This section will cover the automated deployment process and the corresponding Jenkins pipeline to automate the entire process in the ABC Insurance UiPath CoE. Let's discuss the details in the following section.

Automated deployment process in the ABC Insurance UiPath COE

The automated deployment process will be triggered once the UiPath developers complete the code review and merge the code to the development branch. Once the right scripts and resources are in place, the UiPath administrator will merge the code, which will be the trigger for the Jenkins process. The Jenkins pipeline will build, deploy, and run tests in the test environment. Then, once the test results are validated, the script is merged into the UAT branch. This is the second deployment kick-off and the package is promoted to UAT. The UAT test is performed in this environment, and once it is validated, the package is deployed to the production environment.

Automated tests are executed in both UAT and production environments as well. Once the business stakeholders provide the product validation, the UiPath administrator merges the script to the **Production** branch of GitHub. The developers can pull in the latest script from this branch for any future process changes.

The process will be terminated if any of the validation fails. The process is also integrated with ServiceNow to update the status of the change request tickets automatically and email notifications are sent to the right stakeholders.

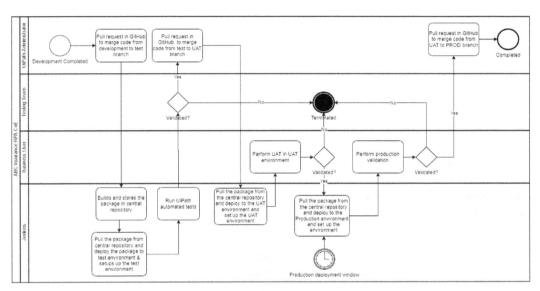

Figure 6.18 – The ABC Insurance UiPath automated deployment process

Next, let's see how this process is automated with the Jenkins pipeline.

The Jenkins pipeline in action in the ABC Insurance UiPath CoE

The main purpose of the Jenkins pipeline is to orchestrate the automated deployment process right from the build phase until the final deployment validation. The different Jenkins plugin features discussed in various sections of this chapter will be used in a sequence to orchestrate this DevOps flow. This mature Jenkins pipeline project is integrated into GitHub and ServiceNow.

The entire pipeline is configured to orchestrate not just a type of UiPath deployment, but that same pipeline is used for all the deployments performed by the ABC Insurance UiPath CoE.

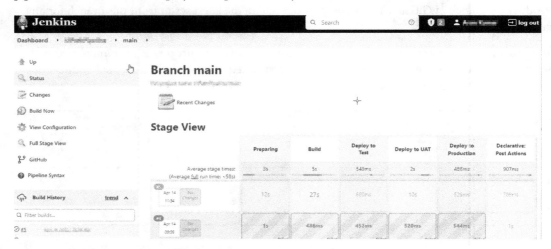

Figure 6.19 – The ABC Insurance UiPath automated deployment process

The initial build is triggered after the preparation phase, where the integration with GitHub is made. Once the NuGet package is uploaded to a centralized repository, then the package is automatically deployed to the **Test** environment.

Automated tests are executed in the **Test** environment if the automated systems tests pass. Once the tests are validated, the script gets promoted to the UAT branch. This action will promote the same package to the **UAT** environment. Preconfigured automated UAT tests are performed in UAT environment The results are passed to a UAT validator, and once the results are verified, the package will be deployed to the **Production** environment during the production deployment window. After the production deployment is done, regression tests will be executed, and results will be passed to the business validators. The business validators will provide the final validation, the entire journey can be monitored, and audit trails are available.

The UiPath scripts maintained in GitHub ensure configuration management and version control; automated tests ensure the quality of the product and automated deployment confirms that the deployment is free of human errors. The post-deployment mail with steps to set up the continuous monitoring will be kickstarted.

Support roles and best practices in troubleshooting DevOps issues

Now that Jennifer is familiar with the current automated deployment setup and all the aspects concerning the DevOps lifecycle in the ABC Insurance Corporation's UiPath CoE, she can play an active role in troubleshooting UiPath DevOps issues.

The Jenkins pipeline has different stages, such as build, test, and deployment; if a DevOps process fails at a particular phase, it can easily be identified from the Jenkins console. Once the build with issues is identified, then the console output logs can be used to dive deep and identify the root cause of the issue.

There are many scenarios where she will be requested to support a typical UiPath DevOps issue, and here are a few best practices and tips to keep in mind when supporting them:

- **Scenario 1**: GitHub commits, and merges cannot be made by the UiPath developer.

  - **Support role**: The UiPath support personnel or administrator needs to understand the GitHub repository branching strategies and rules. The two most common root causes are a) that the developer might not have access to the relevant repository and b) that there is a violation of an existing GitHub branching rule.

  - **Best practice**:

    - Try replicating the scenario from UiPath Studio before moving to the next step. Involve the GitHub administrator to understand any changes to the repository policies.

    - Create a knowledge base of all the GitHub-related issues that UiPath developers can access.

- **Scenario 2**: The build process is failing.

  - **Support role**: The UiPath support or administrator needs to understand the UiPath NuGet package information and why the file is not available in the respective folder. There are two main root causes: either a) Jenkins might not have access to the repository, and GitHub access and credentials may be restricted, or b) the build package version might already be available.

  - **Best practice**:

    - Understand how GitHub and Jenkins credentials are configured and the steps to update the values.

    - Understand the versioning options in the build process.

- **Scenario 3**: We cannot execute UiPath test sets.

  - **Support role**: The UiPath support or administrator needs to understand how UiPath tests are set up in Orchestrator. There are two main root causes: a) that test sets and cases are not configured properly in Orchestrator, or b) that access and resource to test robots are not configured in the respective environments.

  - **Best practice**:

    - Understand how test automation is set up in general.

    - Discuss test requirements with the UiPath Test automation developer and have a prerequisite and operational documents to run these tests successfully.

- **Scenario 4**: Deployments to the environment are failing.

  - **Support role**: The UiPath support or administrator needs to understand why the NuGet package is not deployed to the respective environment. There are two main root causes: a) that UiPath Orchestrator configuration input parameters are not properly provided, or b) that UiPath Orchestrator access is denied.

  - **Best practice**:

    - Understand all the UiPath Orchestrator API keys and respective parameters required to perform the deployment to any environment.

    - Understand the requisite prerequisites, such as the build process, approval, and GitHub merges that need to happen.

These are just some pointers, and there are many possible root causes. Solutions may be very complex in some cases. The main idea of this section is to give students some flavor of what to expect when supporting a UiPath DevOps issue.

> **Note**
> Azure DevOps and GitHub actions are two other popular DevOps options supported by UiPath. It is worth spending time to understand how they are used in different enterprises.

We have walked Jennifer through understanding how the DevOps concept is used in the UiPath program. It's time to recap what we've learned in the following section.

Summary

DevOps is one of the fastest-growing software methodologies, and it's adopted in almost every enterprise. Many UiPath customers have adopted DevOps for their UiPath program in recent times. The main idea of this chapter was to equip the UiPath support team and administrators with a quick introduction to DevOps and its applicability to UiPath programs. This chapter also provided an overall picture of how DevOps is applied to the UiPath CoE. This information will give of UiPath Support trainees a head start in understanding the issues and possible fixes for UiPath DevOps issues. Any new UiPath support personnel member such as Jennifer needs to understand every DevOps phase and have some basic understanding of Jenkins, one of the most popular UiPath automated deployment tools.

DevOps is a vast topic, and there are many books written on DevOps adoption and its supporting platforms. There may be many instances in this chapter where the student needs to research and self-learn to start being able to support UiPath DevOps issues.

This chapter has explained how manual build and deployments were performed in the ABC Insurance UiPath CoE. This section has also covered the process, artifacts, and roles involved in executing this whole manual build and deployment process. It is always good to know the past before we plan. This knowledge helps us to appreciate the value DevOps could bring to the UiPath CoE.

The main DevOps concept and its lifecycle phases were introduced in the following section. Most of the chapter covered how the ABC Insurance UiPath CoE implemented and benefited from these DevOps phases. In the next section, Continuous Development and Continuous Integration phases were explained with the help of UiPath Studio integration to GitHub. How the UiPath Jenkins plugin can be used to perform the build step was also explained. The Jenkins pipeline was also covered to give a flavor of how enterprise-grade UiPath build processes are executed.

Continuous testing was covered next. UiPath Test Suite was briefly introduced and how Jenkins can trigger automated tests was also explained. Continuous deployment concepts were covered in the next section and the different aspects of using the Jenkins plugin with UiPath for deployments were explained in this section. Continuous monitoring is one of the most relevant and important phases of UiPath Support. A different phase of continuous monitoring was detailed in this section, and we will dive deeper into more details in the next chapter.

Finally, the current DevOps process in play in the ABC Insurance UiPath CoE was detailed, as was the role of UiPath Support in this process, along with the discussion of a few best practices.

A UiPath process called Claims work assignment was used as a sample case to explain how the process moves through the DevOps phase from development to deployment in various sections of the chapter to help Jennifer understand the practicality of the concepts. Along with this, best practices were also outlined in each section to add more context to performing these support tasks in the best possible way.

I hope this chapter helped Jennifer get a quick overview of performing UiPath DevOps support activities. Let's discuss the details of UiPath monitoring in the next chapter.

7
Monitoring and Reporting in UiPath

The UiPath monitoring and reporting capability is critical for the overall success of the UiPath CoE program, and it is one of the best enablers for scaling up the entire UiPath automation program. Monitoring UiPath resources such as robots, machines, and jobs, are part of the daily routine in the life of a UiPath Support and monitoring team member. Hence, this chapter is relevant to any new UiPath Support team member and student.

As discussed in the previous chapter, continuous monitoring is an important DevOps phase. This chapter will cover all the aspects of the continuous monitoring approach that are followed at the ABC Insurance Corporation UiPath CoE. We will start by providing an overview of the UiPath monitoring framework. Then, the next sections will cover other monitoring and reporting types at various framework levels, such as business, application, and infrastructure. This chapter also covers different monitoring and reporting tool options and sheds light on some best practices for the UiPath monitoring team in enabling this capability.

The UiPath platform provides three out-of-the-box monitoring features:

- Monitoring in UiPath Orchestrator
- Webhooks and email notifications in UiPath Orchestrator
- UiPath Insights

This chapter will cover all these options in detail. In addition to these features, the UiPath Process Mining offering can also be used for enhanced continuous process monitoring requests, and the details on this will be covered in *Chapter 9*.

At the ABC Insurance UiPath CoE, a dedicated UiPath Support and monitoring team is available to help with continuous monitoring requests and troubleshooting issues. Hence, all the members must go through the UiPath monitoring and reporting framework and the setup training to understand the existing setup. We will explain to Jennifer how various UiPath monitoring and reporting framework components are implemented at the ABC Insurance Corporation UiPath CoE and emphasize the importance of the most critical capability for scaling and maturing the UiPath programs.

> **Note**
>
> In the ABC Insurance Corporation, few enterprise monitoring solutions (such as Splunk and Prometheus) and dashboard solutions (such as Tableau and Kibana) are available. This can be the case in many mature organizations; hence, it is critical for the UiPath CoE leadership to decide on the right mix of monitoring and reporting solutions for their UiPath program. The UiPath Support and monitoring team will work with different external teams to support the unified monitoring solution for the UiPath platform. It is recommended to stick to the enterprise monitoring solution, which can easily be integrated with the UiPath platform without a lot of data transformation or building custom connectors.

Here is what you will learn as part of this chapter:

- Get an overview UiPath monitoring and reporting components and the framework

- Understand business-, application-, and infrastructure-level monitoring and reporting for the UiPath program

- Learn other tools and options available for UiPath monitoring and reporting

- Understand how to support the team members involved in handling monitoring requests

- Learn the best practices in performing these UiPath monitoring and reporting activities

Let's understand the UiPath monitoring and reporting framework in the first section of this chapter.

The UiPath monitoring and reporting framework

Monitoring and reporting capabilities are crucial for maintaining and scaling up any UiPath program. It is important to have a defined policy, roles, processes, and tools for enabling these capabilities to cater to key stakeholders' needs. These details are defined in the monitoring and reporting framework, which is reviewed and approved by a central governance body. The framework components and details are continuously updated to fit the needs of the stakeholders.

Some of the key benefits of having a UiPath monitoring and reporting framework are as follows:

- Improved visibility of the overall UiPath program

- Reduced time spent responding to bot and platform issues

- Generalized processes and tools to support monitoring requests
- Improved UiPath customer satisfaction

> **Note**
>
> Different companies follow two major types of UiPath monitoring and reporting frameworks:
>
> A) A centralized approach, where all the UiPath projects must comply with one common approach. A reduced turnaround time and standardized governance policies and processes are the major benefits of this approach.
>
> B) A decentralized approach, where the UiPath project or function decides how to monitor and report framework components and tools. A reduced total cost of ownership and less maintenance effort are some of the benefits of this approach.
>
> At the ABC Insurance UiPath CoE, the centralized UiPath monitoring and reporting framework is followed.

New UiPath Monitoring team members must understand the framework to excel in their monitoring roles. Getting an overall picture will enable them to handle challenging situations as well.

In the next section, let's discuss the monitoring and reporting framework and its components implemented by the ABC Insurance UiPath CoE team.

The ABC Insurance UiPath monitoring and reporting framework

At the ABC Insurance UiPath CoE, there is a three-layer monitoring and reporting framework to provide continuous monitoring and reporting capabilities for the UiPath program:

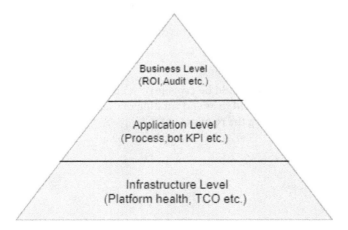

Figure 7.1 – The ABC Insurance UiPath CoE monitoring and reporting framework

Each of these layers has different governing policies, stakeholders, tools, and processes. We will dive deeper into the details of these three layers later in the chapter.

> **Note**
>
> Business- and application-level monitoring and reporting components are within the UiPath CoE scope. Still, in many organizations, the UiPath Monitoring team needs to coordinate with the existing enterprise infrastructure team to define the components of the infrastructure level.

There are different roles involved in executing the monitoring task. Let's look at them in the next section.

The UiPath Monitoring team's roles and responsibilities

ABC Insurance Corporation's UiPath team is a group dedicated to handling UiPath monitoring and reporting requests. There will be a monitoring lead who will manage the operational aspects of the UiPath monitoring and reporting team. This role will report to the UiPath Support and monitoring manager. The monitoring team is involved in various activities, and a few of them are listed here:

- Monitor the successful operational activities of UiPath resources such as Robots, jobs, and machines
- Verify that all failure events are logged and notifications are sent to the respective stakeholders
- Use monitoring dashboards and reporting capabilities to generate request reports and data points
- Help troubleshoot production issues by providing data and logs to enable the support team to handle the request
- Work with development and deployment teams to ensure monitoring and reporting requirements for the new processes are implemented during the release of changes
- Proactively observe data trends and patterns and raise problem tickets for the bot support team

> **Note**
>
> The UiPath Bot monitoring team should be capable of supporting requests at all three levels of the framework. It is recommended to attend training on the monitoring solutions in place in the company before performing the requests.

Now that Jennifer has understood the high-level details of the UiPath monitoring and reporting framework and the responsibilities of the monitoring team members, in the following section, we will touch upon the business-level monitoring and reporting details.

Level 1 – Business-level monitoring and reporting

The first level of the monitoring framework deals with monitoring and reporting business **Key Performance Indicators (KPIs)** that are relevant to tracking the value (ROI) and compliance (SLA) of the UiPath applications.

At ABC Insurance Corporation's UiPath CoE, the UiPath Insights application is used as a primary RPA analytics application for setting up the dashboards, alerts, and reports at this level. The UiPath Monitoring team will set up and support these UiPath Insights dashboards once the applications are in production.

Quantifying the value of returns through real-time ROI dashboards, the validation of compliance to agreed SLA, and complete visibility of the abstracted data points to the functional level of operational indicators of the UiPath applications are some benefits of enabling business-level monitoring. This will provide confidence to business stakeholders to continue investing in supporting existing bots and scaling up the automation programs.

Before we get into the details of the monitoring dashboards, let's take an overview of UiPath Insights.

UiPath Insights

Insights is an RPA analytics offering from UiPath that can track UiPath Robots and application performance, calculate business ROI, bring operational transparency, and help with real-time reports.

A few Insights dashboard templates are available in this application out of the box. UiPath Monitoring team members can use these templates as a starting point to learn about and build customized dashboards that fit the business analytics requirements.

Insights intakes data from the UiPath Orchestrator database, and the support will also be part of the existing arrangements between clients and UiPath. This way, reliance on and risk of failure from external analytics tools such as Kibana and Tableau are eliminated.

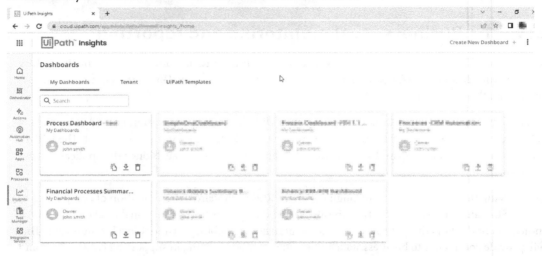

Figure 7.2 – UiPath Insights

New dashboards can be created, or customization of the dashboards is possible from the explore panel, where various data visualizations, filters, and settings are available to support the creation of UiPath dashboards. Once created, the dashboards need to be tested and approved before they can be shared with the stakeholders.

Different roles such as UiPath Insights **Admin**, **Designer**, **Viewer**, **ROI Editor**, and **ROI Viewer** can be configured and used to perform different operations on this analytics application:

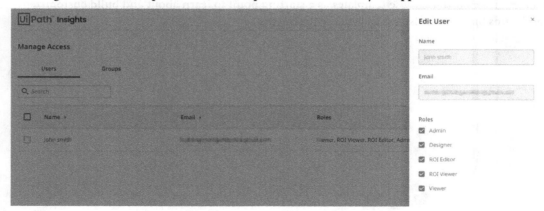

Figure 7.3 – The roles in UiPath Insights

> **Note**
> Proper governance that covers the segregation of duties, naming conventions, and ROI data points needs to be in place before rolling UiPath Insights out into the production environment.

UiPath Insights training is available on UiPath Academy and all UiPath Monitoring team members are recommended to enroll in this training.

Monitoring at the business level

ROI and SLA compliance are the two major indicators monitored at the business level. UiPath Insights has an out-of-the-box ROI dashboard that can be used to monitor the UiPath ROI. An ROI dataset interface can facilitate tracking the manual time and hourly costs involved in executing the process manually.

Once the ROI information is entered, and the filtering criteria are set up, the ROI calculation is performed automatically based on the data from UiPath Orchestrator:

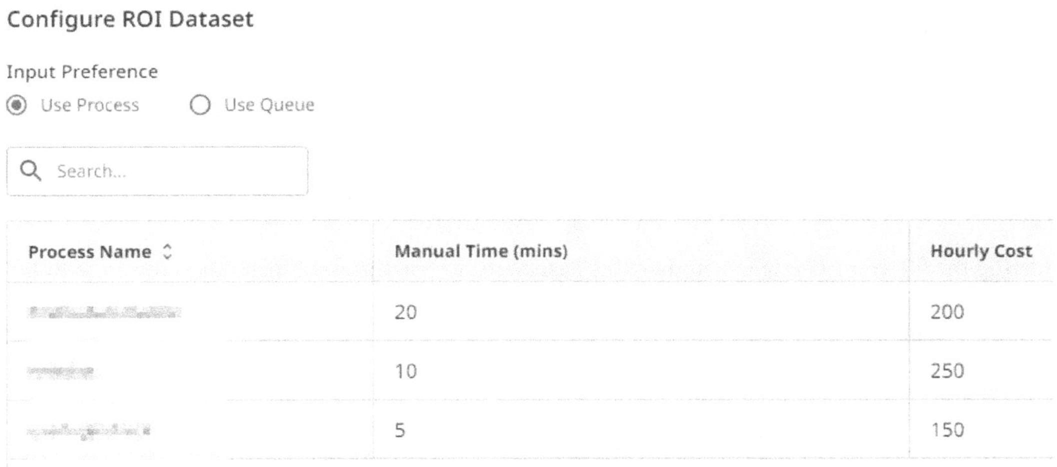

Configure ROI Dataset

Input Preference
◉ Use Process   ○ Use Queue

Q Search...

Process Name ↕	Manual Time (mins)	Hourly Cost
	20	200
	10	250
	5	150

Figure 7.4 – The configuration of ROI

The Finance Process ROI dashboard displays the total time and dollar amount saved on the UiPath processes compared with the baseline data entered in the ROI configuration. This information will be vital for business stakeholders to justify the investments in the UiPath program and to invest additional resources to improve the ROI of the automated compounded processes:

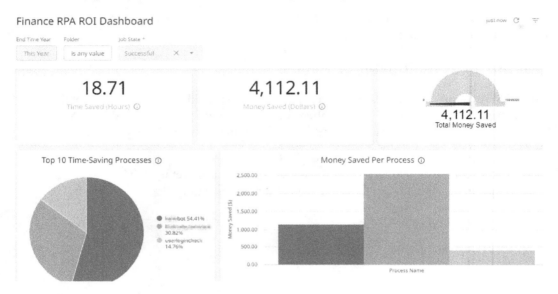

Figure 7.5 – The Finance RPA ROI Dashboard

Updating the filtering criteria will update this dashboard in real time and instantly refresh the ROI dashboard with the right ROI data. Customization to this dashboard is also possible and UiPath Monitoring team members must understand the intricacies of this dashboard.

Alerts can also be set up to send a notification email to the business stakeholders if a particular anomaly is deducted. For instance, if the success rate of a business-critical bot is zero over a long period, then the bot needs attention. As it has financial implications, the business stakeholders are also notified so that a manual contingency plan can be implemented until the bot issues are solved.

Reporting at the business level

UiPath Insights is also used to generate real-time reports out of the configured dashboards. It is recommended that all the reports are automated, eliminating the risk of human error affecting the reported indicators.

The frequency, format, and stakeholders of these UiPath Insights reports can be configured. Filtering and other advanced options are also available for custom and complex reporting requests:

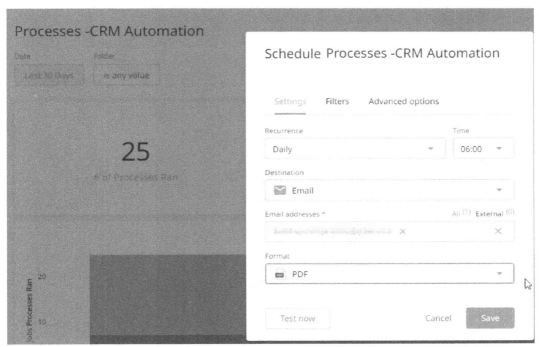

Figure 7.6 – The bot process report configuration

For instance, the SLAs for all the UiPath financial processes at ABC Insurance Corporation are captured in a dashboard. The agreed SLA for the success rate of the UiPath bot process determines that it should be greater than 70% in a week. A report can be scheduled at the end of the week to validate whether the bots meet the agreed SLAs.

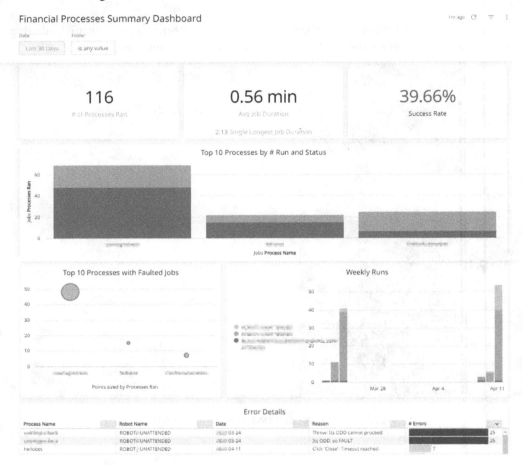

Figure 7.7 – The financial bot process report

Based on the data in the financial dashboard, if the success rate is less than 70%, it is a breach of contract and business stakeholders need to contact their UiPath leadership counterparts to check why the SLA was breached and the next steps to take to rectify the SLA compliance issues.

> **Note**
>
> ROI and process overviews are the two major business reports that stakeholders are very interested in and concerned about at ABC Insurance Corporation. The business-level dashboards and reports are usually standardized at the leadership level, and any changes must go through an approval process.

Having all the business KPIs in reports can be used to present a great business case for continued investment in the UiPath RPA program. Many business KPI-specific UiPath dashboards and reports can be created by the UiPath Monitoring team based on the business requirements.

Now that Jennifer has understood the UiPath business-level monitoring and reporting components, the following section will start with application-level monitoring and reporting details.

Level 2 – Application-level monitoring and reporting

The second level of the monitoring framework deals with monitoring and reporting the details of the UiPath application deployed in the production environment.

At ABC Insurance Corporation, the UiPath CoE, UiPath Insights, and UiPath Orchestrator (monitoring, notification, and webhooks) features are also primarily used for application monitoring and reporting. Hence, a good understanding of these three components is necessary for the UiPath Monitoring team.

Application-level monitoring and reporting of UiPath applications are very useful for the UiPath Support team to get complete oversight of the health of the UiPath applications and be notified if there are any issues. The UiPath CoE leadership team also uses this level of UiPath application health dashboards and reports to maintain and validate the value proposition of supporting these production applications.

Monitoring at the application level

At the ABC Insurance Corporation UiPath CoE, all the UiPath applications are monitored with the help of individual UiPath process dashboards. The dashboard will contain the UiPath application details such as the total number of transactions, success and failure rates, time, and individual transaction details.

The most helpful feature of these dashboards is various filtering options based on time and date ranges, applications supported, and volume. When the application is moved to production, these dashboards are usually created by the UiPath Monitoring team.

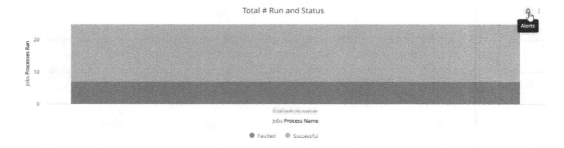

Figure 7.8 – The process status

Custom alerts can be set up on these dashboards to notify the UiPath Support team of unexpected patterns within UiPath application behaviors. For instance, an alert can be generated on a particular UiPath application when there are more than five faulted jobs in an hour:

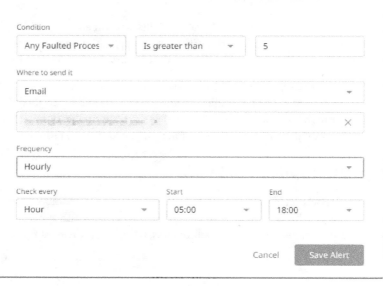

Figure 7.9 – A process alert

Multiple alerts can be set up and the notification channels and participants can be controlled based on the business needs agreed upon with the relevant stakeholders.

> **Note**
>
> Application monitoring dashboard data can be consolidated based on application or business groups. Let's say the finance tenant has a specialized UiPath Support and monitoring team and customized dashboards. This tenant segregation will help reduce the time taken by the UiPath Monitoring team to check individual dashboards one at a time and help fast-track the troubleshooting activities.

In the next chapter, let's try to understand the monitoring feature provided by UiPath Orchestrator.

Monitoring in UiPath Orchestrator

The **Monitoring** tab in UiPath Orchestrator is an out-of-the-box monitoring capability that can be leveraged to get an overview of the UiPath applications deployed in that Orchestrator environment.

The job's success and failure details, machine status, process information, queues, SLA, and license information can be monitored with the out-of-the-box dashboards available in UiPath Orchestrator:

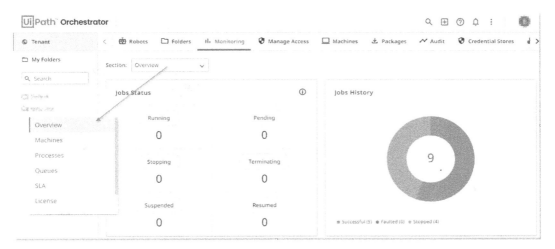

Figure 7.10 – The Monitoring Overview

At the ABC Insurance UiPath CoE, Orchestrator dashboards are constantly monitored every hour, and any anomalies flagged in this dashboard are used to flag potential issues. This is followed by a deep-dive investigation of individual process dashboards.

Email notification and webhooks alerting in UiPath Orchestrator

Email notifications based on a secure SMTP connection are configured when UiPath Orchestrator is set up (**Settings | Mail**). Frequent email notifications about faulted job statuses are shared with the UiPath Monitoring team members. These emails will contain the details on the jobs that get faulted in a robot and the monitoring team member can use this information to analyze the faulted jobs.

Webhooks is an integration feature for external applications. They are termed reverse APIs that send events from a web application to an HTTP collector. This is another out-of-the-box feature available in UiPath Orchestrator that can be configured to send UiPath events out to external monitoring tools to enhance the UiPath application monitoring capability:

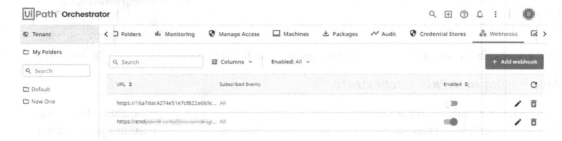

Figure 7.11 – Webhooks in UiPath Orchestrator

Once an external HTTP collector is set up in the enterprise monitoring tool, the collector URL is used to create a new webhook. Different events about jobs, Robots, and queues can be subscribed to, and the event data will be transmitted to the external application collector.

For instance, if a job is stopped or created, the respective UiPath Event data gets transmitted to the collector in JSON format, which will have different parameters that will be useful for the UiPath Support team.

#	Time	Message
1	05/27/2022 10:45:47.053 PM -0500	View as JSON {"Type":"job.stopped","EventId":"763269959","Timestamp":"2022-05-28T03:45:47.053","Jobs": [{"Id":112021383,"Key":"e1ccca51-3b02-4316-b29a-fb3a3978d028","State":"Stopped","StartTime":"2022-05-28T03:44:36.577Z","EndTime":"2022-05-28T03:45:47.0282419Z","Info":"Job Hello was stopped (forced)!\r\n\r\nRemoteException wrapping System.Exception: Job Hello was stopped (forced)! \n","OutputArguments":null,"Robot": {"Id":865166,"Name":"robot2- ","MachineName":"EC2AMA 939F"},"Release": {"Id":601738,"Key":"3a47e37c-0a44-4212-8743-bcc0ab467762","ProcessKey":" "}}],"TenantId":4752,"OrganizationUnitId":3456808} Host:40.112.64.166 ▾ Name:Http Input ▾ Category:uipath ▾

2	05/27/2022 10:44:36.237 PM -0500	View as Raw
		{
		  Type: "job.created",
		  EventId: "763268583",
		  Timestamp: "2022-05-28T03:44:36.237",
		  StartInfo: ▸ {...},
		  Jobs: ▸ [...],
		  TenantId: 4752,
		  OrganizationUnitId: 3456808,
		  UserId: 5821
		}
		
		Host:40.127.192.244 ▾ Name:Http Input ▾ Category:uipath ▾

3	05/27/2022 10:44:36.006 PM -0500	View as Raw { Type: "job.started",

Figure 7.12 – UiPath Events in JSON

Custom email or SMS notifications can be created and triggered based on the business rules defined by the UiPath Monitoring team. Let's say a business stakeholder manually stops a job. A notification will be sent to the UiPath Support team and one of the UiPath Support team members will reach out to the business stakeholders to understand why the UiPath job was stopped and will take the next steps to revert the jobs to a successful run state.

Figure 7.13 – UiPath custom alerts based on webhooks events

Multiple webhooks can be created for different setups of event data transmitted to the external monitoring tools. This additional capability will enhance the UiPath application monitoring and can be used to reduce application downtime.

> **Note**
>
> Internal mail notification on faulted jobs can trigger an automation that can work as a self-heal to faulted jobs. For instance, if the job faults for intermittent issues such as loss of internet connectivity, the faulted job can be automatically retriggered without the need for the UiPath Support team. We will discuss this interesting topic more in *Chapter 9*.

Reporting at the application level

UiPath application-level reports are used by the UiPath CoE leadership and the UiPath Support team to constantly improve UiPath application performance indicators such as availability and reliability.

Reports can be scheduled from the existing dashboards and are usually set up when the UiPath application is onboarded into the production environment. There are different reports on UiPath application monitoring configured in ABC Insurance Corporation's UiPath team. Let's discuss a few here:

- A daily report on the UiPath application and machine health status is very helpful for the UiPath Support team, as they will use this data to fine-tune the application and machines to increase the value of the UiPath applications in future runs.

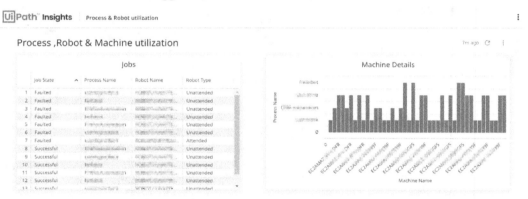

Figure 7.14 – An application and machine snapshot report

> **Note**
> It is good to have the UiPath application reporting requirements documented and signed off as part of the UiPath Support handover checklist.

- A Robot utilization report is an important indicator for the UiPath leadership and support team. This report can be scheduled to run every month and the pattern will help the UiPath Support team schedule and use the Robots available in the production environment optimally. These reports can also be used to see the split in utilization between **Attended** and **Unattended** Robots in a production environment. These insights can be used to make an informed licensing decision.

For instance, if an unattended bot has a higher error rate than other Robots running the same process, it is then a clear indicator that a Robot-specific issue is persistent, and the UiPath Support team needs to investigate and find the root cause of these errors.

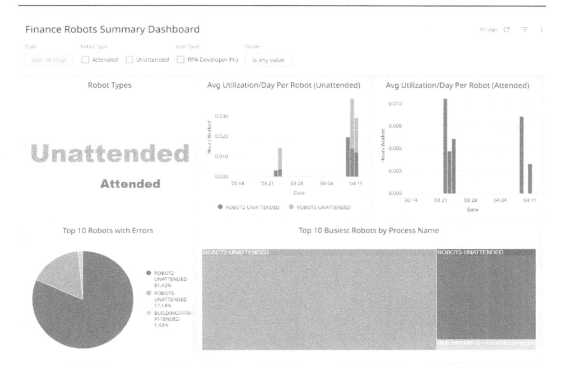

Figure 7.15 – A Robots report for the finance tenant

These are just a few reports explaining the capability of application-level reporting from UiPath Insights. Still, several options are available to produce customized reporting to help the UiPath Support team during application production issues.

> **Note**
>
> The UiPath Monitoring team needs to ensure the Robot licenses are utilized optimally, as this is one of the KPIs closely tracked by the UiPath leadership team.

Now that Jennifer has understood the UiPath application-level monitoring and reporting components, in the following section, we will start with the details of infrastructure-level monitoring and reporting.

Level 3 – Infrastructure-level monitoring and reporting

The third level of the monitoring framework deals with monitoring and reporting on the infrastructure details (i.e., virtual machines, databases, application servers, etc.) that needed to be set up for the UiPath RPA platform.

At ABC Insurance Corporation, a centralized infrastructure monitoring team is assigned to support all the enterprise application needs. The UiPath Monitoring team works closely with the centralized team to create dashboards and reports to report on the health of the UiPath platform infrastructure.

The infrastructure monitoring on the enterprise resources is necessary only if either Orchestrator or the Robot are hosted on-prem. The following matrix is a reference guide for involving UiPath Monitoring within infrastructure monitoring needs:

Orchestrator	Robots	Studio	Infrastructure Monitor
On-Perm Cloud	On-Perm Cloud	On-Perm Cloud	Yes
UiPath Cloud	On-Perm Cloud	On-Perm Cloud	Yes
UiPath Cloud	UiPath Cloud	On-Perm Cloud	No
UiPath Cloud	UiPath Cloud	UiPath Cloud	No

Figure 7.16 – An infrastructure monitoring decision matrix

> **Note**
>
> At ABC Insurance Corporation, UiPath Orchestrator is hosted on the UiPath cloud. Still, the Robots' virtual machines are hosted on-prem cloud instances, hence the UiPath Monitoring team only has to get involved in monitoring the health of virtual machines. If Orchestrator is hosted on-prem, then the application servers and databases must additionally be scoped into infrastructure monitoring.

Reducing the downtimes of the UiPath platform, increasing the availability of Robots, providing preemptive alerts to avoid any potential issues (e.g., CPU load, disk space, etc.), and having visibility into all the UiPath infrastructure costs are some of the advantages for the UiPath Monitoring team members.

Let's understand some more details on monitoring tasks at the infrastructure level.

Monitoring at the infrastructure level

Most cloud infrastructure providers have monitoring solutions that can be leveraged for infrastructure monitoring. There are also specialized monitoring tools such as Splunk, Dynatrace, and Datadog available that you can enable with enterprise-level infrastructure.

The enterprise monitoring team needs to work with different application monitoring teams such as the UiPath platform to set up the working and governance process.

Any UiPath Monitoring team member needs to understand these governance processes to perform their regular infrastructure monitoring activities.

Monitoring tools

A monitoring tool will enable the creation of custom dashboards to monitor different infrastructure health indicators. In the ABC Insurance UiPath team, Splunk is the primary tool of choice.

One of the most important dashboards is the CPU utilization of all the servers, especially the Orchestrator and database servers. The following graph shows the CPU spike at different time intervals in a day:

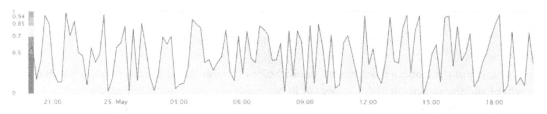

Figure 7.17 – A UiPath Orchestrator server showing CPU utilization

If UiPath Orchestrator is hosted on-prem, this dashboard will be the first checkpoint in case of application issues. For instance, if the CPU utilization has increased due to the peak volume of UiPath jobs, this may have a performance impact. It is recommended to have an alert turned on to notify of abnormal CPU spikes.

The UiPath Monitoring team should be able to detect anomalies from these patterns of spikes and take preventative action before it impacts the whole platform performance or even crashes it.

Figure 7.18 – UiPath Robot machine availability

The availability of the Robot's machine is a good choice of dashboard that can correlate the Robot throughput and the agreed-on SLA. Outages of Robot machines will directly impact the UiPath jobs and increase the failure rate. Hence, regular monitoring of this dashboard is needed.

The memory and disk space of UiPath infrastructure resources are also very important metrics that need to be monitored with these custom dashboards.

The basic infrastructure monitoring is built on application availability, memory, disk space, and CPU utilization. More advanced infrastructure monitoring dashboards can be configured based on need.

> **Note**
>
> It is recommended to have a unified UiPath platform-specific dashboard that has all the infrastructure components listed. This will be useful for the UiPath Monitoring team to have an overview of the infrastructure, and it will also save time and reduce risk in the UiPath platform.

At the ABC Insurance UiPath CoE, custom monitoring dashboards that reflect the availability of CPU, memory, and disk space on UiPath Robot virtual machines built with Splunk are in place, and hence, Jennifer has to go through this basic infrastructure training to perform these monitoring activities.

In the following section, let's cover the reports generated from infrastructure monitoring.

Reporting at the infrastructure level

Reports on the UiPath platform infrastructure are important at both the UiPath CoE leadership level and operational level. Automated reports generated from the infrastructure monitoring tool can be set up on both scheduled and real-time triggers.

Abnormal resource utilization such as CPU spikes, memory leaks, and disk space issues will trigger individual notifications on resources. A report on the affected resources is automated, generated, and shared with the respective stakeholders if a particular threshold is breached. These real-time reports can be used as a benchmark to check whether the platform's health is restored to its normal state.

The application or Robot response can be tracked in real time and the performance issues can be flagged in real time. A report is triggered when a UiPath application response time does not meet the agreed SLA.

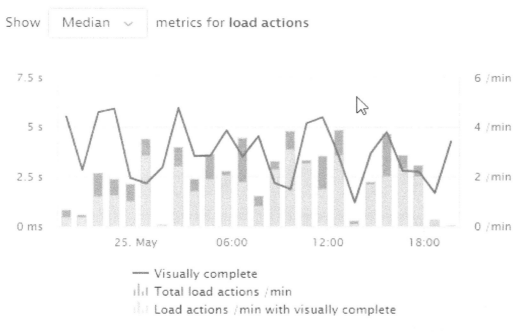

Figure 7.19 – The UiPath target application response time with a Robot

Real-time reports can be used to not only quickly troubleshoot production issues but also to validate the fix. These are just sample real-time reports and there will be different reporting requests based on the need of the stakeholders.

These are a few traditional scheduled reports that are shared with the relevant stakeholders within an agreed time frame:

- UiPath Robot machine health:

  - UiPath platform resource health is an important aspect to translate into a report that will hold information on the state of the core infrastructure needed to execute the UiPath active jobs. This report can automatically be generated from the enterprise monitoring tool.

  - The following is a sample image of a report that shows the AWS EC2 server instances that are active and offline during problematic days of a month:

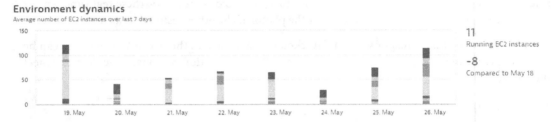

Figure 7.20 – A snapshot of the UiPath platform for EC2 servers

- Target business application health:

  - It is equally important to report the health of the target automated business application. When bots start to work actively on a target business application such as SAP or CRM, the health indicators that point to performance are monitored. Having a healthy target system will make sure that the ROI promised by the bot programs is reached.

> **Note**
>
> If UiPath Orchestrator is hosted on-prem, it is very important to get a regular health status for the databases hosting the UiPath data. Generating database health reports will reduce the risk of UiPath platform failure.

These are just a few samples of infrastructure reports generated by the ABC Insurance UiPath team. This just the tip of the iceberg; more sophisticated reports can be automated based on need.

Businesses can use these reports to do the following:

- Validate the TCO and ROI for the UiPath program
- Get a snapshot of the infrastructure health and validate the costs
- Check the UiPath application's availability, capacity, and reliability

The UiPath Monitoring and Reporting team at ABC Insurance Corporation have a catalog of infrastructure reports that is generated periodically. The UiPath monitoring lead will check whether all the activity reports are generated and shared with the relevant stakeholders.

> **Note**
>
> The UiPath monitoring lead should set up a periodic review of the data points on the reports that need to be scheduled with the relevant stakeholders, and action items that need to be defined to rectify the concerns highlighted in the reports.

Now that Jennifer has understood all three of the levels of UiPath monitoring and reporting components, in the following section, we will discuss other popular tool options and best practices.

Monitoring and reporting options and best practices for the UiPath program

As previously discussed in this chapter, there are many options for supporting the monitoring and reporting of the three layers of the monitoring framework.

Popular monitoring and reporting tool options

We discussed using UiPath Insights for Level 1 and 2 monitoring and reporting needs, but there are a few other popular options:

- **Elasticsearch and a Kibana dashboard**: Elasticsearch is a distributed open search and analytics engine that can ingest UiPath logs and enable quicker search capabilities. Kibana is a data visualization tool that works with Elasticsearch and can be used to create the dashboards used for UiPath monitoring.

- **A reporting database and Tableau**: The custom log from the Robot executions is written to a reporting database. Tableau is used to create dashboards and reports by connecting to this custom UiPath logs database. This option is popular when limited monitoring needs to be enabled on only critical processes.

Splunk is a very popular application used for Level 3 infrastructure monitoring. Apart from Splunk, there are other major players in the enterprise monitoring space: Datadog, New Relic, Dynatrace, App Dynamics, and SolarWinds are all active within enterprise-level monitoring. These applications can be used to monitor the UiPath infrastructure and build advanced analytics solutions.

Next, let's understand the details of the UiPath CoE reporting.

UiPath program and CoE reporting

Apart from the three levels of reporting, the monitoring and reporting team also helps prepare operational reports on the UiPath program or CoE.

One of the major indicators is the bot support indicator for various processes in the bot support team oversight. The support team needs to track the operational health of the bots in production, and frequent reporting on the status will benefit from tracking the value proposition of the whole program.

The UiPath development and release delivery reports are designed to report the performance of the UiPath development and release teams. Different metrics such as delivery velocity, defects, and release stories can give the leadership team a detailed view of the current performance of the bot development teams.

The support team's performance can also be reported, covering metrics such as the total number of support tickets handled, SLA breaches, tools, and automation to improve bot support. Metrics on the bot support team performance will give confidence to the leadership team to plan for UiPath CoE scalability.

There may be ad hoc reporting requests such as monthly trends on ticket volume, bot health, and team performance. The UiPath Monitoring team will be the team responsible for helping generate these reports on demand.

> **Note**
> Having a good reporting setup in place will ensure the success of the UiPath CoE.

Now that Jennifer has understood the importance of UiPath program reporting, let's try to understand the best practices.

Best practices in handling monitoring and reporting requests

UiPath Monitoring team members need to have a best practices handbook, which needs to be frequently updated. Here are a few best practices and tips to keep in mind.

- Business-level monitoring and reporting:

    - ROI and SLA data cannot be updated without proper authorization from both the business and UiPath CoE leadership team

    - A list of business stakeholders to report mapping needs to be maintained

    - Dashboards have to be tested in lower environments before they are promoted to production

- Application-level monitoring and reporting:

  - Documentation on business requirements, UiPath Insights dashboards, and reports should be available in a centralized repository

  - Proper naming conventions must be followed for all the dashboards and reports

  - Customizing these dashboards needs to be centralized and managed so that the default dashboards and reports are not affected

- Infrastructure-level monitoring and reporting:

  - Upskilling for the enterprise monitoring tools with training and demos will be helpful

  - Creating architectural viewpoint diagrams and having all the resources documented will be helpful for troubleshooting issues quickly

  - Open communication on infrastructure changes such as upgrades and patching exercises will be needed to improve the availability

These are just some pointers for the UiPath Monitoring team member to quickly excel in their assigned monitoring tasks.

> **Note**
>
> The UiPath Process Mining solution can also be leveraged to provide advanced continuous monitoring of end-to-end processes. UiPath bots are also used along with other solutions such as BPM and ERP to complete a business transaction. The UiPath Support and monitoring team needs to understand this aspect as well. The details of this advanced capability will be covered in *Chapter 9*.

We have walked Jennifer through and understood all the monitoring and reporting components and applications used in the UiPath program. It's time to recap what we have learned in the following section.

Summary

Monitoring and reporting are important for all aspects of business management, and the same holds for UiPath operations. UiPath monitoring and reporting is one of the most important capabilities for supporting a UiPath CoE or program. Many UiPath customers ignore the benefits that this capability brings to the RPA program until the entire program is in a bad state.

Understanding this capability's importance within all levels of the UiPath program is necessary for a successful program. The UiPath monitoring and reporting framework provides a structured way for new UiPath Monitoring team members to quickly understand different stakeholders' viewpoints in the UiPath RPA program from a monitoring and reporting needs perspective.

This chapter provided a high-level overview of the UiPath monitoring and reporting framework followed by the ABC Insurance UiPath Monitoring team. The roles and responsibilities of UiPath Monitoring team members were then discussed. The following section covered the introduction of UiPath Insights and its high-level capabilities, before diving into the details of Level 1, business-level monitoring.

The monitoring and reporting solution around UiPath CoE ROI and SLA compliance were primarily detailed in the following section. We then covered different ways of and tools for constructing application-level monitoring and reporting for the UiPath platform.

This was followed by a discussion on infrastructure-level monitoring and reporting for UiPath platforms, and various options were covered. Popular external enterprise monitoring tool options were also covered in this section.

Finally, UiPath's internal UiPath CoE reporting needs and a few best practices for the UiPath Monitoring team were also discussed. The main goal of this chapter was to provide a 360-degree view of various monitoring and reporting options for the UiPath platform. Using the information on various dashboards, alerts, and reports, the monitoring team members can then support the UiPath Monitoring team requests. Creating a UiPath process dashboard, checking the health of UiPath virtual machines (if there are any outages), and adding or adjusting ROI data for a new process in business dashboards are a few examples of monitoring requests that the reader needs to be prepared to approach.

There are many ways to make UiPath monitoring and reporting smarter with the latest technologies and custom utilities, reducing manual effort and moving towards an automated approach. These details will be covered in the final chapter, *Chapter 9*.

The overall big picture will help the monitoring team member excel in their daily UiPath monitoring and reporting tasks. I hope this chapter helped Jennifer get a quick overview of performing the relevant UiPath monitoring and reporting activities.

This concludes the second section, *Part 2, Administration, Support, DevOps, and Monitoring in Action*, of the book. In the next chapter, we will move on to the final section, *Part 3, UiPath Maintenance and Future Trends*, starting with the details of the UiPath platform's maintenance and upgrades.

Most of the topics of the next chapter will be relevant to the on-prem UiPath setup and they will be mentioned in the subtitle of the topics.

Part 3: UiPath Maintenance and Future Trends

In this part, you will get an overview of UiPath platform maintenance, performing a UiPath upgrade for an on-premises scenario, and enterprise UiPath platform reporting. In addition to this, you will also learn about the support role in risk management, IT security, auditing, how to support a multi-node Orchestrator setup, disaster recovery, enterprise non-core UiPath platform support activities, and future trends in UiPath support and administration.

This section contains the following chapters:

- *Chapter 8, UiPath Maintenance and Upgrade*
- *Chapter 9, UiPath Support– Advanced Topics and Future Trends*

8

UiPath Maintenance and Upgrade

Periodic health checks on various aspects of a system are referred to as system maintenance, and updates to various system components are referred to as upgrades. Like any other enterprise IT platform, maintaining and upgrading the UiPath platform is vital to the success of the UiPath program deliverables and the value proposition promised to the stakeholders.

The contents of this chapter will be relevant and helpful if one or more UiPath core components such as Orchestrator, Robots, and Studio are hosted in an on-premises environment. The UiPath leadership team can use this decision matrix to include maintenance and upgrade tasks training for their UiPath Support team members:

Orchestrator	Robots	Studio	Maintenance & Upgrade
On-Perm Cloud	On-Perm Cloud	On-Perm Cloud	Yes
UiPath Cloud	On-Perm Cloud	On-Perm Cloud	Yes
UiPath Cloud	UiPath Cloud	On-Perm Cloud	Yes
UiPath Cloud	UiPath Cloud	UiPath Cloud	No

Figure 8.1 – UiPath maintenance decision matrix

This chapter will provide an overview of UiPath RPA platform maintenance and upgrade activities. We will start by covering regular platform maintenance activities such as database maintenance and then touch upon topics such as RPA infrastructure and platform maintenance. This chapter will also cover various UiPath upgrade scenarios, which will be helpful for the UiPath Support and platform administration team.

In ABC Insurance Corporation, Robots and Studio are hosted on-premises; hence, maintaining and upgrading the Robots and Studio components is necessary. UiPath's Support and monitoring team is available to help continuously maintain and upgrade these two components. Thus, all the members must go through the UiPath maintenance and upgrade training to understand various scheduled and ad hoc tasks involved in maintaining and upgrading UiPath Robots and Studio. We will explain the overall UiPath maintenance and upgrade tasks in various sections of this chapter and emphasize the importance of performing them while covering their subsequent impact on the success of the UiPath programs.

> **Note**
> The bulk of this chapter's maintenance and upgrade tasks will revolve around UiPath Orchestrator as it is the core and complex product for maintaining and upgrading in an on-premises environment. As ABC Insurance Corporation uses a Cloud UiPath Orchestrator and an on-premises cloud version of Studio and Robots, only a subset of this chapter was relevant to Jennifer. Therefore, will not use this persona in each section of this chapter.

In this chapter, we will cover the following topics:

- UiPath platform maintenance (on-premises)
- UiPath system health check (on-premises)
- Robot machine maintenance
- Other UiPath maintenance activities
- UiPath platform upgrade activities (on-premises)

First, let's understand on-premises UiPath platform maintenance.

UiPath platform maintenance (on-premises)

The two main components of any core UiPath platform, be it UiPath Orchestrator or UiPath Test Manager, are as follows:

- The application server
- Database

The enterprise infrastructure team usually maintains application servers, and database servers are managed by the Enterprise **Database Administration (DBA)** team.

As a UiPath Support team member, it is vital to know the basic maintenance activities that are performed in these infrastructure components of the UiPath platform as it will help fast-track root cause analysis and issues with the UiPath platform:

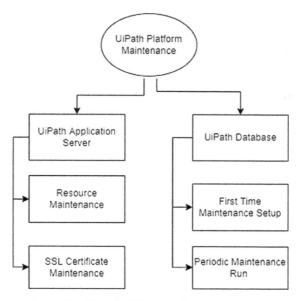

Figure 8.2 – Sample UiPath platform maintenance

New UiPath monitoring team members must understand the details of the UiPath application and database servers. They will have to monitor the health of these infrastructure resources and support maintenance activities on these servers if needed. Getting an overall picture will enable them to handle challenging situations as well.

> **Note**
>
> Other regular maintenance activities that are performed on these servers by the infrastructure team, such as database index reviews, security scans, and more, might be outside the scope for UiPath Support team members.

In the next section, we'll discuss UiPath application server maintenance details.

UiPath application server maintenance

The application server in the UiPath context is a web server used for hosting UiPath Orchestrator, Test Manager, and more as web applications. The application can be controlled using the **Internet Information Services** (**IIS**) available on Windows Server.

If multiple node Orchestrators have been set up, we will need to have equivalent instances of these web servers available to support them. As these servers hosting the UiPath applications are the core of the entire UiPath platform, certain activities are allocated to UiPath Support team members. Now, let's discuss the resource maintenance task.

Resource maintenance

Having optimal resource utilization such as CPU, memory, and disk space for the web server is needed for the steady-state operation of the UiPath application. For instance, any unexpected CPU spikes for a prolonged period will surely affect the performance of UiPath Orchestrator. In turn, the bot operations and jobs may fail as the connections between the robots and Orchestrator will be lost.

Overutilization of disk space and memory will also lead to similar performance issues on the web server. To avoid the issues, it is recommended to periodically check the application server's resource health.

> **Note**
>
> As discussed in *Chapter 7*, an enterprise monitoring tool such as Splunk can be utilized to build dashboards to track these metrics and add alerts to notify the concerned team if there is a breach of a resource utilization agreement.

Another common maintenance activity the UiPath Support team must be aware of on an on-premises setup is checking for web certificate validity.

Secure Sockets Layer (SSL) certificate maintenance

A **Secure Socket Layer** (**SSL**) is needed to certify that the web server's connection is secure from the browser. It is commonly termed **web certificates**. Three types of web certificates are accepted by the IIS Manager:

- Trusted certificates, such as those from Go Daddy, Verizon, and others
- Domain certification – valid in an environment, such as the ABC bank
- Self-sign – for non-domain users.

It is recommended that a valid certificate is installed on the UiPath application server with an unexpired validated date:

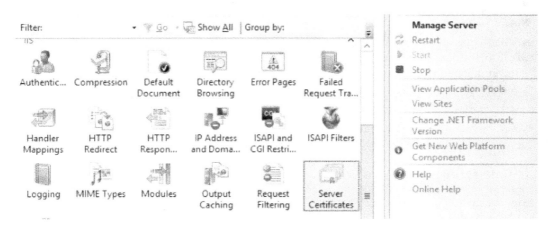

Figure 8.3 – Sample IIS and Server certificates

These certificates need to be renewed once the expiration date is reached. Users will not be able to log into UiPath Orchestrator without a trusted certificate valid through the current operation date.

It is important for the UiPath Support team to closely monitor the certificate expiry dates in case the infrastructure team is not tracking the expiry date.

> **Note**
>
> The server certificate thumbprint, for example, `a9575985dcd4f513f9614c15c986944 c6abfed23`, available on the **Certificate Details** tab, needs to be updated in the `appsettings. json` file (`C:\Program Files (x86)\UiPath\Orchestrator\Identityfile`) in the `Identity` directory to establish a relationship between UiPath Orchestrator and the web certificate.

Now, let's discuss UiPath database server maintenance.

UiPath database server maintenance

MS SQL is the recommended database for hosting UiPath on-premises applications, and just like any other database, the UiPath product team recommends a few scheduled maintenance activities.

A UiPath database consists of several tables for storing the details of users, jobs, robots, and more. As the UiPath operation grows, the data gets accumulated, and the size of the database keeps expanding. One possible solution is to archive historical data in an archival database and purge it from the primary database.

> **Note**
>
> A UiPath upgrade cannot be completed if there are huge database sizes, say greater than 50 GB; if so, the data migration will fail during the on-premises UiPath Orchestrator upgrade. It is recommended to archive and clean up data from different targeted tables, as discussed in this section.

UiPath recommends that UiPath maintenance be set up once the application is set up. Hence, we can be sure the database's size is under the maintainable size recommended by the UiPath team.

First-time setup

A series of activities must be performed to set up UiPath database maintenance. These activities need to be planned with the help of the DBA, who has administrator rights on the UiPath database:

1. Identify the tables that take up the most space in the database, such as logs, RobotLogs, QueueItems, jobs, audit logs, and more.

2. Get consensus and approval on the data retention timeline – for example, 90 days. The timeline will determine the amount of historical data that can persist in the primary database at a given time.

3. If a lower environment such as the development or test environment is available, then a separate database instance for that environment will exist. Set up the database maintenance in that lower environment to test the queries or stored procedures.

4. Create an archival database, such as UiPathOrchestratorArchives, and the tables chosen to be archived and prefix them with Archive to identify them later – for example, ArchiveLogs for a logs table, ArchiveQueueItem for QueueItems, and so on.

5. Stop the UiPath application, such as Orchestrator, from the IIS server and make sure all current transactions are completed.

6. Execute the queries to Archive (insert into) to move the data from the primary tables, such as logs, into ArchiveLogs, and then purge (delete) the data from logs. This step must be repeated for all the tables chosen in *step 1*.

> **Note**
>
> Some sample queries you can use to set up the database tables are available at https://docs.uipath.com/installation-and-upgrade/docs/maintenance-considerations.

7. Verify that more data was created than the retention data (90 days) and that it's been removed from the primary databases tables, and that the same data is available in the archival database.

8. Once the initial checks have been completed, start Orchestrator from IIS in the application server and run a few sanity tests on the lower environment orchestrator. Monitor the successful operational activities of UiPath resources such as robots, jobs, machines, and more.

9. Benchmark the size of the UiPath database for future reference.

10. Once the test environment database maintenance is successful, perform *step 4* to *step 9* on the production database.

Best Practice

If new database tables have been identified and chosen to be archived, then the queries can be added to this existing list in the future.

Once the initial database maintenance is completed, it is necessary to schedule the query execution regularly, say every month or week, depending on the volume of data generated by the UiPath application. In the next section, we'll discuss scheduling these tasks as SQL server jobs.

Periodic maintenance setup

The main idea of database maintenance is optimal UiPath platform performance. Hence, UiPath database maintenance needs to happen periodically. The following are a few guidelines for setting up this activity:

*   Calculate the volume of data generated by the UiPath application on the primary database. This is an important criterion to determine the frequency of executing these database maintenance jobs. Let's say the UiPath CoE leadership decides to execute this database maintenance activity monthly.

*   The UiPath Support team needs to work with the Enterprise DBA team to create mechanisms to schedule these jobs periodically.

*   Each query is run as a separate scheduled SQL job, and the job sequence is also determined before the schedule is in place.

*   Do a trial run on the lower environment and see how long it takes for the archival and purging activities to be completed.

*   Notifications for both successful and failure job runs are set up.

*   Move the scheduled SQL jobs to be run against the production database.

It is recommended to stop the UiPath application during this scheduled maintenance window. Otherwise, the SQL jobs may fail to archive and purge data.

Once the UiPath database maintenance has been set up, it will reduce a lot of effort from the UiPath Support and DBA teams to optimize the UiPath platform's performance.

> **Note**
> Job scheduling applications such as Redwood RunMyJobs or active batches can also perform this scheduled query or SQL job run. Notification emails can be configured to have an oversight of the data being archived on a regular interval.

Now that we understand the high-level details of UiPath platform maintenance and the responsibilities of the support team members, let's look at the different UiPath system health checks.

UiPath system health check (on-premises)

The UiPath system comprises three core elements:

- Application
- Data
- Infrastructure

It is the responsibility of the UiPath Support team to get the health status and plan regular maintenance activities on each of the layers:

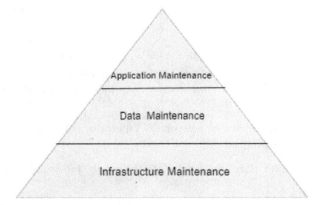

Figure 8.4 – UiPath system maintenance

Different health checks are designed to get a report on each of the three blocks. A respective mitigation activity follows every system health check, which are the regular maintenance activities that are performed by the UiPath Support team.

We'll get an overview of UiPath application maintenance in the next section.

UiPath application maintenance

As many UiPath processes get onboarded continuously to a **Business As Usual** (**BAU**) or support mode, the process performance may be affected in the long run. Hence, the UiPath Support team is responsible for performing a few planned maintenance activities on long-running UiPath processes. A few sample activities have been listed in this section to give you a basic understanding of this type of maintenance activity:

- **Volume distribution between bots**: It is good to check the volume, failures, and SLA compliance data of different UiPath processes and see any correlation to the bots they assigned. Optimizing the UiPath processes that are run on different bots is recommended to ensure that all the volumes are evenly distributed between bots and make the assignments appropriately:

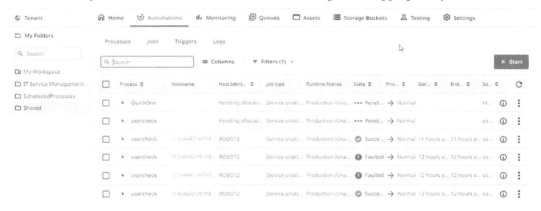

Figure 8.5 – UiPath Orchestrator jobs

- **Check scheduled runs**: The UiPath Support team must check on all the triggers and validate the information with relevant stakeholders. There may be instances where a few triggers might be disabled or changed by mistake. This will be a good opportunity to rectify these mistakes before business stakeholders escalate them.

- **Bot accesses are renewed**: Bot access to certain business applications, shared drives, or licensed products must be checked and updated. This maintenance activity will reduce the bot's downtime because of access and license issues.

> **Note**
>
> The UiPath Support team needs to work with the UiPath developers to understand UiPath process-specific maintenance tasks and document them as part of the support handover document.

Now, let's discuss maintenance activities at the data layer.

UiPath data maintenance

The Enterprise data management team generally handles maintenance activities in the data layer of the UiPath application. Database integrity checks, optimization, index reorganization, running update statistics, and performing a full database backup are a few of the DBA team's common database activities.

The UiPath Support team can monitor the health of the UiPath database after a scheduled maintenance activity using custom dashboards from the Enterprise infrastructure monitoring team to ensure it does not affect the UiPath application's performance:

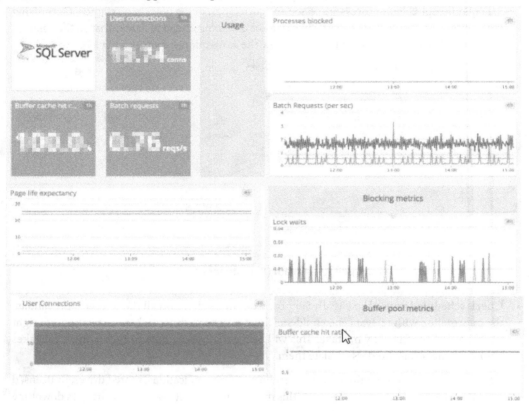

Figure 8.6 – Custom dashboard for the database

> **Best Practice**
>
> It is recommended that the DBA team shares the planned maintenance cycle of the UiPath database servers with the UiPath Support team in advance.

Now, let's discuss maintenance activities at the infrastructure layer.

UiPath infrastructure maintenance

Periodic health checks need to be planned for the UiPath infrastructure. Health reports of different UiPath servers, virtual desktops, and other components such as HAA and the load balancers maintained by the Enterprise monitoring team need to be reviewed by the UiPath Support team and reported back to the respective teams.

For instance, if there are CPU or memory issues on a particular infrastructure component after Windows patching, then action items need to be planned by the UiPath Support team to resolve the issues. This could involve restarting the servers after patching them.

A few of the recommendations for the UiPath Support team regarding maintaining the UiPath infrastructure are as follows:

- **Web SSL certificate:** The UiPath application server's SSL certificate expiry needs to be checked by the UiPath administrators regularly every 6 months to track the validity of these SSL certificates.

- **Windows patching:** All Windows-based machines need to be patched with a Windows update to minimize IT security compliance issues. The enterprise infrastructure team usually controls the policy of security updates.

> **Best Practice**
>
> Information on regular Windows upgrades or security patching exercises on these core UiPath application servers must be shared with the UiPath Support team so that the application downtime can be planned and communicated to stakeholders.

After every Windows patching exercise, it is important to check the UiPath infrastructure's health. The UiPath Support team can automatically develop an automation utility to set up customized alerts on escalation issues automatically. For instance, if CPU utilization was 100% on the UiPath Orchestrator server for more than 3 hours, then Orchestrator event logs, the application health of currently running jobs, and more are collected and then shared with the UiPath Support team.

- **Data backups:** UiPath administrators must ensure that all the UiPath infrastructure component images are backed up regularly by an automatic batch program.

> **Best Practice**
>
> It is recommended that both the UiPath application and database server's images and data are backed up every night during an agreed schedule. The Enterprise infrastructure team needs to be involved in this backup activity. These images will be used for disaster recovery scenarios.

Now that we understand the different health checks and mitigations that are performed by the UiPath Support team, let's understand the robot machine maintenance activities.

Robot machine maintenance

Robot machines are also enterprise infrastructure resources that are usually managed by the Enterprise infrastructure team. As digital workers or bots access these machines to perform critical business operations, the UiPath Support team must oversee the robot machine maintenance activities by working closely with the enterprise infrastructure team.

As discussed in *Chapter 1, Understanding UiPath Platform Constructs and Setup*, there are different ways to configure robot machines. Virtual desktops and servers (for high density) are two popular robot hosting options in this section's scope.

In ABC Insurance Corporation, each UiPath robot is hosted on individual AWS workspaces (cloud-based virtual desktops). Hence, a good understanding of robot machine maintenance is necessary for the UiPath Support team.

> **Best Practice**
>
> Server options such as AWS EC2 can also be used to host multiple robot accounts on a single machine. The UiPath robots hosted in these server environments are called **high-density robots**, and similar robot machine maintenance activities need to be performed.

Robot machine health check

As the enterprise infrastructure team manages robot machines, the UiPath Support team needs to work with them to set up some dashboards and alert them about some of the recommended health checks on the robot machines in place at ABC Insurance Corporation.

These health checks are deep dive checks that will not just report on machine availability but will go into a few details that need to be validated periodically to ensure the machines are available for bot operations:

- **Disk space**: The disk space on a virtual robot machine needs to be maintained for the effective performance of the UiPath robot, and dashboards and alerts for breaches in expected limits need to be communicated – for example, 70% disk space in `C:\`.

- **Memory**: High memory utilization. For example, if memory utilization is >95%, this is another data point that needs to be tracked on the UiPath robot machine.

- **CPU utilization**: UiPath performance will be directly affected if the CPU utilization is consistently high – that is, it's more than 90%. So, this parameter also needs to be tracked.

- **Software licenses**: Software licenses are needed in robot machines for business applications to complete the transactions – for example, Microsoft Office Suite and Acrobat.

- **Shared drive access**: Transferring files to and from the robot machine is another important prerequisite that needs to be monitored periodically. The robot's access to location can be configured and controlled.

- **The business application accesses**: The business application needs to be accessible from the robot machines for the jobs to run successfully. Hence, it is important to check if the business applications are available and if application connections or sessions can be established from robot machines:

Robot Machines	DiskSpace	Memory	CPU Utilization	Software Licenses	Shared Drive Access	Business Application Access
Robot Machine 1						
Robot Machine 2						
Robot Machine 3						
Robot Machine 4						
Robot Machine 5						

Figure 8.7 – Process status

Other parameters may be necessary, and these can be added to these health check jobs as needed.

Custom alerts can be set up on these parameters to notify the UiPath Support team of unexpected patterns in UiPath robot machine behavior. This could be an alert generated on a particular UiPath machine, a Microsoft Office 365 license has expired, or particular shared drive access was removed.

Multiple alerts can be set up, and the notification channels and participants can be controlled based on the need agreed upon by the relevant stakeholders.

> **Note**
>
> The robot machine checks can be automated and scheduled to run periodically. Critical alerts can also be set up, and custom automation jobs can be created for this activity.

Now, let's try to understand a few details on the precautions and mitigations that should be implemented by the UiPath Support team and the infrastructure team.

Precautions for mitigating robot machine health issues

The UiPath Support team needs to monitor the robot machine dashboards and alerts regularly. SLA and operating procedures must be established to handle robot machine health issues. These are reactive measures that UiPath Support team members need to be involved in.

Apart from these, some proactive measures and mitigations can be implemented to minimize the UiPath robot machine issues. These maintenance jobs need to be designed and operated by the UiPath Support and monitoring team. A few of the maintenance jobs used in ABC Insurance Corporation are as follows:

- **Custom jobs for disk cleanup**: Custom jobs or automation can be built to temporary cleanup files in different drives such as C:\, as well as in application-specific temp such as Outlook temp files.

- **Access and license checks**: A sanity test or canary automation can be designed to test for all access policies compliance. A similar test can be run to check software license utilization.

- **Periodic restarts**: Robot virtual machines need to be restarted regularly. This activity can also be a scheduled maintenance job.

The UiPath Support team needs to have a few mitigation measures in place in case the robot machine is not recoverable. New robot machine instances may need to be spun up on short notice. It is good to have an **Amazon Machine Image** (**AMI**) or any other type of image ready with all the prerequisites for operating the UiPath robot and establishing the connection to UiPath Orchestrator.

Now that we have covered all the robot machine maintenance activities, in the next section, we will look at other UiPath maintenance activities that UiPath Support members usually perform.

Other UiPath maintenance activities

UiPath maintenance is a continuous activity, and different tasks are assigned to the UiPath Support team to perform these activities at scheduled time frames. A few common maintenance activities not covered in the previous sections can be seen in the following diagram

Figure 8.8 – Other maintenance blocks

Let's take a closer look.

Access control

A few access control-related activities are performed by the UiPath Support and Administration teams:

- **UiPath infrastructure user groups**: Regular audits of user groups and respective users who have access to the UiPath infrastructure resources such as the web server and database server need to be completed periodically. User data maintenance needs to be performed by the UiPath Support team with the support of the Enterprise infrastructure team.

- **UiPath application user groups**: Next, the access control maintenance task is scoped for UiPath applications such as Orchestrator, Insights, Test Manager, and so on. User groups and users need to be listed, and updates must be performed periodically by the UiPath administrators.

- **UiPath CoE access**: The UiPath support team also maintains access for the UiPath CoE artifacts and user access that need to be checked periodically. Folder-level access needs to be checked as well, and unauthorized users need to be removed.

Now, let's understand a few activities related to resource optimization.

Resource optimization

A few resource optimization activities are performed by the UiPath support and administration teams:

- **Removing unnecessary files**: UiPath deals with inputs in the form of files dropped in different shared drives, S3, and other locations. The bots generate the audit report, output file, and other files when they perform the business operations. It is necessary to clean up these temp files regularly. If this activity is overseen, it may affect new file creations and stop bot operations.

- **Auto archival for Outlook**: If bots work on email automation, it is recommended to have auto archival turned on so that the email box limit is not breached, and the automation run is not affected.

- **Periodic restarts**: One of the known issues with Windows-based servers and workstations is that they need to be restarted periodically. These restarts must be planned so that bot operations have minimal impact. The Enterprise infrastructure team usually handles this, but the UiPath Support team needs to work with their counterpart to plan the schedules for these restarts.

> **Note**
> The UiPath Support team can develop customized automation utilities to clean up temporary files, provision bots to perform these resource optimization checks, and more. These automation jobs can be scheduled to run regularly to help with maintenance activities.

Now, let's understand a few maintenance activities related to monitoring and reporting.

Monitoring and reporting

The following are a few monitoring and reporting maintenance activities that are performed by the UiPath Support and Administration team:

- **Remove unused dashboards and reports**: UiPath monitoring solutions such as Insights or Kibana dashboards need to be maintained as well. Periodic maintenance activities must be performed to remove dashboards belonging to offboarded UiPath processes. The UiPath Support and Monitoring team must also stop alerts and reports from the processes.

- **Update subscribers**: The business stakeholders of the respective UiPath processes who receive the reports and alerts from the UiPath monitoring tool need to audit periodically. The UiPath Support and monitoring team needs to coordinate with the respective product teams to get the updated business stakeholders to complete this activity.

- **Update ROI data**: The UiPath Insights ROI dashboard is a critical business monitoring tool. It is necessary to cross-check the ROI data that's been configured in the dashboard periodically.

Documentation activities

- **RPA process inventory update**: The UiPath process repository, which contains all the UiPath process lists currently operating in production, needs to be updated periodically. The updated information from these files can be used to fast-track issue resolution. The UiPath Support team must work with the release and product team to keep this information updated periodically.

- **UiPath architecture and infrastructure**: The UiPath architecture and infrastructure document will contain all the UiPath platform and infrastructure-related details such as the application server, robot machine, software license, and more. The UiPath Support team needs to keep updating these documents periodically with the help of the Technical Architecture team.

- **Technical and support documentation**: The UiPath process technical and support troubleshooting guide is another piece of important documentation that will help the UiPath Support team to figure out the root causes quickly and resolve the issues. The UiPath Support team needs to work with the UiPath developers to look for recent changes and new processes released in production and update the relevant documentation.

In ABC Insurance Corporation, the UiPath Support lead is responsible for ensuring the UiPath routine maintenance activities are completed as per schedule, and that audits of these activities are tracked and presented as part of the UiPath CoE reporting.

> **Note**
>
> If Elasticsearch is used to store the logs related to the robot job, then you need to clean up shards periodically for the Kibana dashboards to have optimal performance.

Maintaining the UiPath platform on-premises is a complex activity. A coordinated effort from the UiPath Support and infrastructure teams is needed to perform these tasks and successfully maintain a high-performance UiPath platform.

Now that we have covered most of the UiPath maintenance activities, let's understand the details of the UiPath platform upgrades.

UiPath platform upgrade activities (on-premises)

UiPath applications are constantly updated with new features, and it is recommended to upgrade to the latest version of the UiPath platform to ensure new UiPath capabilities and features are available to the UiPath development and user communities within the enterprise. The upgrade scenarios are only applicable for on-premises installations of UiPath products. UiPath issues **Fast Track Support (FTS)** – that is, x.4; for example, 2021.4 – and **Long-Term Support (LTS)** – that is, x.10; for example, 2021.10. As the name suggests, mainstream support for FTS is just for 6 months, but for LTS, it is for 24 months.

The FTS version is usually used for evaluation purposes and for performing a pilot study of the new features, as well as for enterprises to upgrade to the LTS version of UiPath products.

The release notes of the UiPath product need to be referred to if you wish to figure out the correct version of the UiPath product that needs to be upgraded.

The UiPath Support and Administration team is usually involved in coordinating the on-premises upgrades. UiPath developers, testers, and business users are also involved in validating the platform once the upgrade is completed.

> **Best Practice**
>
> The UiPath platform installer and MSIs can be requested from UiPath through a support ticket. It is recommended to download and keep the installation files in the respective machine's local folders to save time during the upgrade.

In the next section, we'll learn about the UiPath Orchestrator upgrade.

UiPath Orchestrator upgrade

An on-premises Orchestrator upgrade is a major activity for which the UiPath Support and Administration team are involved on a periodic calendar. If there are multiple UiPath Orchestrator environments, then the lower environments such as the test or UAT environments need to be upgraded before upgrading the production Orchestrator environment.

The following diagram shows the high-level UiPath Orchestrator upgrade phases. These are just guidelines for the UiPath Support team to perform the UiPath Orchestrator upgrade:

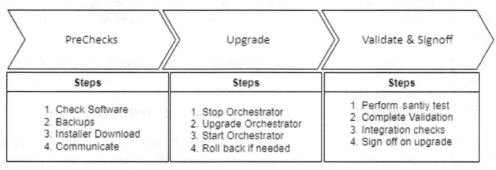

Figure 8.9 – UiPath Orchestrator upgrade

First, we'll look at the prechecks.

Prechecks

Follow these steps:

1. Use the UiPath Orchestrator documentation to understand the software requirements for the latest version – for example, NET Framework or ASP.NET module. Database size and web certificate validation are additional checks that must be checked before moving forward.

> **Note**
>
> The UiPath platform configuration tool introduced during the on-premises installation can also be used before upgrading the Orchestrator. This utility can be downloaded from `https://docs.uipath.com/installation-and-upgrade/docs/platform-configuration-tool`.

2. Backups need to happen before the upgrade is performed:

    I.   Create a Windows Server image to host the UiPath Orchestrator with all the drive data.

    II. Create a UiPath Database to host the Orchestrator data.

3. Download the installer for the target version of Orchestrator by raising a request with the UiPath team.

4. Send the official communication to all stakeholders on the downtime during the UiPath Orchestrator upgrade.

Upgrade

Follow these steps:

1. Once the backups are completed, the UiPath Orchestrator must be stopped from the IIS.

2. Run the installer in an elevated administrator mode; for example, `UiPathOrchestrator.msi /lvx* upgrade.log`. If multiple nodes are used, then we need the output parameter that needs to be used in secondary nodes – for example, `UiPathOrchestrator.msi OUTPUT_PARAMTERS_FILE = "C:\backup\upgradeparams.JSON /lvx* upgrade.log` (`/lvx*` will make sure the upgrade logs are generated).

> **Best Practice**
>
> For a multimode Orchestrator, use `PARAMETERS_FILE` – for example, `upgradeparams.json` – from the primary node set up. You must use two additional parameters when running the `installer.msi` file: an upgrade pass with `SECONDARY_NODE =1` and `PARAMETERS_FILE=upgradeprams.json`; for example, `UiPathOrchestrator.msi SECONDARY_NODE =1 PARAMETERS_FILE=upgradeprams.json /lvx* upgradenode2.log /Q` (`Q= Quiet mode`).

3. Once the upgrade has been completed successfully, we can start the UiPath Orchestrator node in IIS and verify the upgrade has been completed successfully from the upgrade logs.

4. Roll back. An automatic rollback happens if the installer faces any errors during installation. If the application is not accessible or the data is corrupted, we need to use the Windows AMI and database to roll back forcibly to the previous version

Validation and sign off

Follow these steps:

1. **Perform a sanity check**: Log into Orchestrator and check whether all the basic features and data (packages, user, and so on) are intact after the upgrade. Run a few test jobs and see if the UiPath robots can execute them and that the logs are being generated.

2. **Do a detailed validation**: Check the user and group policies and data, as well as all the major features of Orchestrator such as triggers, queues, notifications, and more. Enable all the triggers and execute multiple jobs. Try to validate all the critical bot processes and involve business users and QA for their validation.

3. **Integration checks**: Perform validation on jobs that are triggered from external applications, webhooks, Test Suite and Manager, UiPath Insights, or external monitoring applications.

4. **Sign off**: Once all the validation is completed, sign off from the business user's UiPath platform upgrade team and support team. At this point, the UiPath Orchestrator enters hypercare mode. Communication regarding the successful upgrade and the platform being online can be shared with all the stakeholders.

> **Note**
>
> Windows server administrator privileges are needed to run the installer on the Orchestrator nodes; hence, the enterprise infrastructure team must be invited during this upgrade exercise. It is also advisable to have DBA administrator privileges for the UiPath database server in case of troubleshooting or rollback scenarios.

It is recommended to upgrade Studio and Robots when the Orchestrator is upgraded so that the entire platform upgrade can be completed in a single stretch by the UiPath administrators. Let's try to understand how to upgrade UiPath Studio.

Studio upgrade

UiPath Studio and its flavors, such as Studio Pro and StudioX, can be upgraded with the help of the UiPath Studio MSI. The Studio upgrade, validation, and rollback steps will be covered in this section.

Upgrade

Follow these steps:

1. Run the latest version of the UiPath Studio MSI installer in Administrator mode.
2. Ensure the license is activated by connecting to the Orchestrator through the attended robot session.

Validation

Follow these steps:

1. Check if you can switch profiles – that is, to UiPath Studio Pro, UiPath Studio, or UiPath StudioX.
2. Log into Studio, create the same workflows, and publish them to the UiPath Test Orchestrator or a local drive.

Roll back

Follow these steps:

1. Uninstall the current version of UiPath Studio.
2. Install the previous version of UiPath Studio.
3. Validate if you can create workflows and publish to Orchestrator or locally.

> **Best Practice**
> UiPath development needs to be paused during the UiPath Studio upgrade; hence, the Support team needs to coordinate with the development leads to plan this exercise so that the impact on the project is minimized.

Next, let's learn how to upgrade robots.

Robot upgrade

Like any other UiPath application, robots can be upgraded with the last version of the MSI installer. The robot upgrade, validation, and rollback steps will be detailed in this section.

Upgrade

Follow these steps:

1. Stop the UiPath Robot and UiPath RobotJS Service in the Windows services panel.
2. Run the Robot MSI installer and let the installation complete.
3. Restart the UiPath Robot and UiPath RobotJS Service in the Windows services panel.
4. Check the connection between the Orchestrator and the robot, as well as the system tray.
5. Perform these steps on all the robot machines.

Validation

Follow these steps:

1. Run a test job on a single robot from the Orchestrator to check if the job is working as expected.
2. If that sanity test passes, then run the same sample job on all the upgraded robot machines.

Roll back

Follow these steps:

1. Stop the UiPath Robot and UiPath RobotJS Service in the Windows services panel.
2. Uninstall the current version of the robot from the application wizard.
3. Install the previous version of the robot.
4. Restart the UiPath Robot and UiPath RobotJS Service in the Windows services panel.
5. Run some sanity tests to validate this.

The same procedure must be performed on all the robot machines, which is time-consuming; hence, there may be downtime for the UiPath robots until this exercise is completed. These upgrades are usually planned in phases and during off-working hours for larger robot volumes.

> **Note**
> Administrator privileges are needed to open **Orchestrator settings** to connect the robot to the Orchestrator. Hence, it is recommended to have an enterprise infrastructure administrator during this exercise.

Now, let's learn more about other on-premises UiPath product upgrades.

Other popular on-premises UiPath product upgrades

Apart from the three core UiPath products, there are a few more on-premises UiPath products where the UiPath Support team needs to get involved in helping with the upgrade. A few of these products are mentioned here:

- **Test Manager**: To upgrade a UiPath Test Manager, take a backup of the application server and test manager database server. Run the MSI file to upgrade to the current version and tie it up with the UiPath test environment. The test manager will interact with the test suite module in the UiPath Orchestrator. It is good to upgrade the test manager when the UiPath Test Orchestrator instance is upgraded so that the validation can be performed along with the UiPath Orchestrator upgrade.

- **Insights**: Take a backup of the database server before the Insights upgrade. Run the MSI installer to upgrade and ensure the Orchestrator version is the same as the Insights one. Once the upgrade is completed, check if the user and process data have been migrated properly by logging into different roles to view the dashboards and alerts.

Like the UiPath Support team supports the complete suite of the UiPath platform, it is recommended to understand the upgrade steps. This will be a routine activity and having expertise in these tasks will be essential to sustain and grow in your role as a UiPath Support analyst.

> **Note**
>
> Certain versions UiPath products such as Robots, Studio, and others are incompatible with UiPath Orchestrator. Please refer to the compatibility matrix for more details: `https://docs.uipath.com/overview-guide/docs/compatibility-matrix`.

Now that we understand the different upgrade scenarios, let's understand a few related best practices.

Best practices in UiPath upgrade activities

UiPath monitoring team members need to have a best practices handbook that needs to be frequently updated. Here are a few best practices and tips to keep in mind:

- Orchestrator upgrade:

  - Double check if the server image and database backup have been completed and saved in a safe location. They may be needed in case of rollback.

  - Have a playbook for all the steps of the upgrade activity.

  - If there are lower environments, please perform upgrades before attempting the same in a production environment.

- Have an enterprise downtime of bot operations communication sent out early.
- Have all the support teams, such as DBA, Infrastructure, Tester, and so on during the upgrade exercise.

- Studio, Robots, and other upgrades:
  - Disconnecting the attended robot in the Studio machine during an upgrade is unnecessary
  - Have a phase-wise list of robot upgrade plans to ensure no robot is missed
  - Performing a sanity data check and applying a hypercare period is recommended after any UiPath platform upgrade is completed

With that, we have walked through and understood all the UiPath maintenance and upgrade topics. Now, let's summarize this chapter.

Summary

As many existing UiPath customers are on the on-premises UiPath platform, maintenance activities are mandatory knowledge that needs to be shared with the UiPath Support team. UiPath maintenance activities are not given due importance compared to other support activities. Since UiPath platforms are getting bigger in scope, it is recommended that these activities are highlighted, and new UiPath Support team members are trained.

As ABC Insurance Corporation uses a cloud UiPath Orchestrator and an on-premises cloud version of Studio and Robots, only a subset of this chapter was relevant to Jennifer. Therefore, we did not use this persona in each section of this chapter.

Different UiPath platform maintenance activities are performed on application and database servers. First, we covered IIS, web server certificate, and UiPath Orchestrator database maintenance. Then, we covered important details that are needed for application maintenance, where sample activities were discussed.

Next, we learned where the UiPath Support team needs to coordinate with the DBA and Infrastructure teams to schedule routine maintenance activities such as database backups and Windows patching.

After that, we covered robot machine maintenance. Different maintenance activities need to be performed to achieve optimal robot performance. Then, we covered other maintenance activities, such as access control and resource utilization.

Finally, we covered various UiPath platform upgrade activities for UiPath Orchestrator, Studio, and Robots and briefly touched upon other UiPath product upgrades.

I hope this chapter helped you get a quick overview of UiPath maintenance and upgrade activities. In the final chapter, we'll discuss a few advanced UiPath Support topics and future trends that the UiPath Support and Administration team members need to be aware of.

UiPath Support – Advanced Topics and Future Trends

Due to the highly competitive RPA vendor landscape, UiPath products and its ecosystem are changing rapidly. The automation scope and complexity of the RPA use cases are also increasing due to business needs. Based on these changes, the demand for increased productivity from UiPath support and administration is expected; hence, UiPath support team members must be equipped to support these rapidly evolving platform ecosystems and keep up with the current trends to stay relevant to the business and IT landscape.

This chapter will provide an overview of UiPath RPA's advanced support areas, such as the UiPath self-service catalog. Then, it will introduce how different custom UiPath support and monitoring utilities can be built and utilized to add value to the RPA setup.

Then, you will learn how UiPath personnel are supporting **Business Continuity Planning (BCP)** and **Disaster Recovery (DR)**. Later, UiPath requests related to IT security, risk, and auditing will be covered. This chapter will also highlight how to extend the core support principles to support the extension of UiPath RPA platform components such as Test Manager, Document Understanding, and Process Mining. Finally, this chapter will cover future trends in the UiPath support space such as automated support, containerized deployments, and multi-vendor ecosystems.

Advanced tools, utilities, and standards used in ABC Insurance Corporation's UiPath support team will be discussed in different sections of this chapter. Hence, all the members, including Jennifer, must go through the advanced topics training to understand the inner workings of these advanced capabilities.

In this chapter, we will cover the following topics:

- UiPath Self Service support catalog
- UiPath support apps
- Advanced topics

- UiPath product ecosystem support

- Future trends in UiPath support and administration

First, let's look at an overview of the Self-Service support catalog setup for the ABC Insurance Corporations UiPath support team.

UiPath Self Service support catalog

There is a recent trend of providing end users with features to get their requests addressed with self-service capabilities in the software application support domain. These new capabilities will not only improve customer experience but also reduce the time it takes for the support team to do the tasks. In ABC Insurance Corporation, a similar self-service catalog was hosted in an ITSM system, ServiceNow, where UiPath platform users and customers can get access to many of the support and monitoring services available on the ServiceNow site:

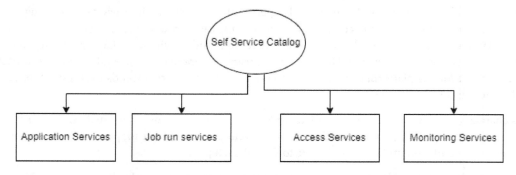

Figure 9.1 – Self Service catalog

Let's try to understand how the self-service catalog was set up by ABC Insurance Corporation's UiPath Support team. We will start by looking at the UiPath application services.

UiPath application services

In ABC Insurance Corporation, all the UiPath product licenses are managed by the UiPath administration team; hence, requests for a new setup or making a change to the UiPath core and supporting system requests would flow through this team before they get redirected to the Enterprise IT support team.

The main advantage of having these services is that they will streamline the request flow and help you do the following:

- Set up new UiPath Studio or Studio Software on developers' machines.

- Set up new UiPath attended or unattended robots on virtual machines.

- Set up new configuration in UiPath Orchestrator such as tenants, folders, and more.

- Set up integration to external systems such as Splunk and Tableau from UiPath.
- Set up other UiPath applications such as test bots, process mining connections, and more.

Next, let's get some details on the job self-services.

UiPath job run services

These automated services relate to UiPath jobs that perform the business processes in a production environment. The main idea is to reduce the UiPath support tickets raised to get the status of a UiPath job or to retrigger failed/missed ones:

- The users are provided access to the view status of a job and are allowed to start/stop a UiPath job that they have permission to access from the self-service portal.
- Access the execution logs of the UiPath jobs. Being able to access the UiPath job's execution logs will bring transparency to the operational job's progress and will reduce support tickets.
- Access the audit reports of the UiPath processes for their jobs in storage drives. The audit or output reports of UiPath jobs will be necessary to understand the details of transactions that were successful or failed.
- Monitor the queues and transactions of their respective processes. For a process with a high volume and tight SLAs, the end users may need to track the pending items in the respective queue.
- View the status of the robot that is executing its process (if applicable). This may be a proactive way to reaffirm to end users that the robots are available to process the planned volume of transactions.

Now, let's look at UiPath access services.

UiPath access services

These automated services relate to the UiPath application and resource access requests raised by stakeholders. The main idea is to reduce the UiPath support tickets raised just to request access to a UiPath product or service:

1. The developers and users request UiPath Studio or Studio X access for citizen development programs. Both internal and external users must raise the request with this option to bring in the visibility of the UiPath product license usage.
2. Business users can access requests to UiPath Task Capture, Test Manager, Action Center, and more.
3. Access UiPath Insights for dashboards or reports. This can be role-specific access, read-only, or edit.

4. Access to shared drives or repos for audit reports or input/output files that host the UiPath process data and reports.

5. Access to the UiPath CoE document repository to browse the available artifacts.

All the access requests will go through the governance approval process before being processed.

Now, let's look at UiPath monitoring services.

UiPath monitoring services

Most of the UiPath monitoring services are raised by business stakeholders. The main idea is to have automated services that can help reduce the UiPath Monitoring team's work:

- Request new dashboards from an existing template in UiPath Insights. Users will be able to choose from the list of available UiPath Insights dashboard templates.

- Request new alerts set on the existing widgets for UiPath dashboards. The name of the widget and alert details need to be provided on the available options page.

- Request new reports from dashboards to be delivered on a set schedule. Changes to existing report schedules are also provided in the same panel.

- Request to add or remove IDs from the distribution list of the reports. The email IDs of users in the distribution list on UiPath Insights reports are generated from existing dashboards.

- Request UiPath Infrastructure health to support their business processes, such as virtual machines, application servers, and more. These reports are generated by the Splunk enterprise monitoring tool.

> **Note**
> It is recommended to have a UiPath program related to **Frequently Asked Questions (FAQs)**, links to training and education, and links to the IT/UiPath support helpdesk on the self-service portal, which will reduce many support requests to the UiPath support team.

With that, Jennifer understands the self-service options that have been built by the ABC Insurance Corporations UiPath support team. Now, let's look at the different UiPath support apps that have been developed by the ABC Insurance Corporation UiPath support team.

UiPath Support apps

Most of the UiPath Support and administration work is manual and takes time. As an automation team, the UiPath Support team can also build a few automation applications (apps) using UiPath or any scripting tool to streamline their regular tasks and reduce the effort required. Four different types of apps are built and used in the ABC Insurance Corporation UiPath support team:

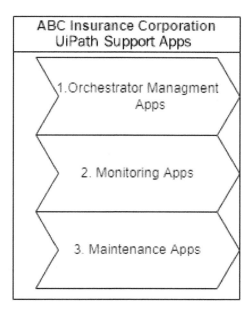

Figure 9.2 – ABC Insurance Corporations UiPath Support apps

Let's start by understanding how the ABC Insurance Corporation UiPath Support team used custom-built apps to manage UiPath Orchestrator.

Orchestrator management apps

Managing and administering UiPath Orchestrator is an important activity for the UiPath Support team and consumes most of their daily work. Identifying a few scenarios that keep recurring and automating the tasks is recommended. This would enable the UiPath Support team members to work more efficiently and spend time troubleshooting complex issues.

A few APIs are available in UiPath Orchestrator that can be used to build apps. In the next section, we'll learn more about these Orchestrator APIs.

UiPath Orchestrator APIs

In simple terms, an **application programming interface** (**API**) is a means for applications to share information. In practical terms, let's assume a client application, such as a banking web page, sends an API request with user authentication details to the database banking server application. Then, the server sends an API response to authorize or deny authentication.

The same concept applies to sharing information or requesting operations on UiPath Orchestrator. UiPath Orchestrator has a defined set of APIs that can be used to do the following:

- Retrieve information using a GET method – for example, get the list of users assigned to the Orchestrator folder

- Add new information using the POST method – for example, to start a new job

- Update existing information using the PUT method – for example, to edit robot information

- Remove information from Orchestrator using the DELETE method – for example, delete a user account

The complete list of API references is available in Swagger, which is available in all UiPath Orchestrator instances and can be accessed using {Orchestrator UR}\orchestrator_/swagger/index.html#/FoldersNavigation:

Figure 9.3 – UiPath Orchestrator API on Swagger

> **Note**
>
> The complete definition of available Orchestrator APIs can be found at https://docs.uipath.com/orchestrator/reference/api-references. Having complete knowledge of the available APIs will enable the UiPath Support team to build apps that can be used for Orchestrator management apps.

Now that we covered the basics, let's try to understand a few sample utilities that are used for Orchestrator management:

- **Provision and unprovision robots**: UiPath robots are associated with a folder that's used to execute processes available in the folder. This is done through the robot group or account assignment on the folder. For instance, during the monthly robot maintenance activity, the robot's access to the folder must be removed. This operation is performed manually by the UiPath support team. Hence, an app was created to automatically add or remove a group of robots to/from the associated folder structure, saving a lot of time for the UiPath support team.

- **Start and stop jobs**: There were many Level 1 requests for the UiPath Support team to start or stop an UiPath job in production based on business stakeholders' requests. Hence, an app was created to do the following:

  a) Start a particular job on a robot or bunch of robots

  b) Stop a single job or multiple jobs

  This app was linked to the self-service portal to reduce business stakeholders from contacting the UiPath support team.

- **Monitor queues**: There were many requests when the business volume of the UiPath production jobs was high during peak seasons for processes. The business stakeholders were constantly requesting the current status of pending jobs in queues to check if the SLAs could be met. Hence, an app was created that monitors the configured queues for the high-volume UiPath processes. Regular notifications can be configured along with the requested queue status, and the requests can be placed via the self-service portal.

- **User and robot access management**: Once the user access has been submitted and approved by the UiPath administrators, the app will automatically add or remove the user(s) or robot account(s) to/from the Orchestrator. 80% of access requests will be handled, freeing up the UiPath support team's time to troubleshoot high-priority items and perform continuous improvements.

Next, let's try to understand how apps are used in different monitoring activities.

Monitoring apps

Monitoring and reporting are a segment where many UiPath administration requests are raised. The scope covers all three levels: business, application, and infrastructure. Let's look at a few sample apps that were developed and used in ABC Insurance Corporation's UiPath support team:

- **Process monitoring**: Automatic solutions can be created to monitor data from the UiPath Orchestrator jobs, queues, and logs to look for any abnormal behavior in operation patterns and flag it. For instance, if the queue is empty for 4 hours, there might be a problem with the upstream system that must be fixed. Another example of an anomaly is a long-running job or logs not generating for a long time.

- **UiPath application and database maintenance checks**: Many maintenance activities were planned for the following:

  I.   UiPath applications such as application password resets, access control updates, and more.

  II.   Database maintenance, such as archiving and purging UiPath data

  These activities need to be monitored manually every week. Hence, an app was created to monitor if the planned maintenance activities were completed on schedule. This app also provides an audit report for future reference.

- **Infrastructure health**: An enterprise monitoring tool (Splunk) is used in ABC Insurance Corporation, where a dashboard was created to track the infrastructure components of the UiPath platform. This includes the following:

  I.   The connection status between all the machines in the UiPath infrastructure

  II.   Performance metrics such as the CPU utilization, network latency, and response time of the UiPath infrastructure components

  III. Capacity metrics such as disk storage and the memory of different machines in the UiPath infrastructure

  Even though the alerts were set up on these dashboards to be triggered when a particular threshold was reached, the alerts needed to be constantly monitored. Hence, an app was created that will consolidate the health issue of the UiPath infrastructure and inform the right stakeholders. They can mitigate the risk and avoid the issue from happening.

- **ROI and daily status reporting**: Many bot process consolidation reports, such as those showing data from different reports, need to be consolidated into a custom template. These were requested by ABC Insurance Corporation's business stakeholders. This included reports showing the ROI data for all CRM processes across all business units or a reconciliation report of failed UiPath jobs across all tenants that need to be restarted.

  Hence, an app was developed that allows you to consolidate or reconcile multiple bot process metrics into customized reports and share them with the relevant stakeholders.

Finally, let's understand how apps are used in different maintenance activities.

Maintenance apps

Maintenance is another area where the UiPath administrator and support team were heavily involved. As most of the activities can be performed by following **Standard Operating Procedures (SOPs)**, few of the activities were shortlisted. The following apps were developed to help the UiPath support team with the planned maintenance activities:

- **Business application access renewals**: UiPath robots access business applications just as a human would and perform business operations on them. Many business application credentials have an expiry date (due to IT security policies), so they need to be renewed. There were 20 business applications by UiPath robots in ABC Insurance Corporation, so it was a time-consuming manual task. Hence, an app was developed that could automatically renew business application access to configured robots on a scheduled date. This app not only saved time but also reduced the downtime of Robots due to access renewal issues that would occur.

• **Automated windows patching downtime**: Another common issue in ABC Insurance Corporation was that there was no support for the windows security patching exercises that happen on all the UiPath Robot machines every month. This is a mandated exercise from IT security that's performed by the infrastructure team. The issue is that the robot's connectivity may be disturbed due to system reboots. Performance issues were also observed.

UiPath support team members are usually assigned to bring down the UiPath robot in sequence just before windows patching is performed in the virtual machine. It was a time-consuming activity that happened over weekends; hence, an app was developed that ensured all jobs were stopped on the robot under maintenance, and the jobs were triggered once the bot machine was available.

• **Robot machine resource maintenance**: In mature UiPath CoE at ABC Insurance Corporation, there were multiple robot machines and various resources such as a shared drive, Outlook, and more that needed to be maintained so that the robots had an optimal performance on those machines.

Manual maintenance activities were performed ad hoc, and a few robot jobs were faulted as there were disk space issues. Hence, the UiPath support team developed an app to run them on a schedule that would free up resources such as audit files, input files, or outlook temp files on the robot machines.

> **Note**
>
> UiPath support and monitoring team members need to be technically sound enough to build these apps independently. That will automatically encourage continuous UiPath support and monitoring team improvements without needing external developers to help.

Now that Jennifer understands the details of different custom apps used by the UiPath Support team, let's look at a few advanced UiPath support activities.

Advanced topics

The UiPath team member's scope of support has increased in recent years as the UiPath CoE has become more mature to scale up the business operations. Many complex and challenging work items have been added to UiPath support teams; we will look at a few in this section. Let's start by discussing disaster recovery and the business continuity plan in the UiPath program.

Disaster recovery and the business continuity plan

Policies, processes, procedures, and tools that enable the continuity of vital services to business operations are part of a **Business Continuity Plan (BCP)**. **Disaster recovery (DR)** refers to the technical aspect of bringing back a faulted application or platform to enable business continuity. In mature enterprises, BCP and DR are defined in an enterprise-wide policy.

The UiPath support and administration teams need to understand the DR and BCP procedures for UiPath-related applications. They should prepare SOPs, processes, and tools to handle these complex scenarios.

BCP for UiPath applications should include steps, stakeholders, and SLA details that must be met in case of outages regarding the UiPath platform or infrastructure. For instance, there was an extended outage of the CRM application in ABC Insurance Corporation. Hence, the UiPath support team activated the BCP for the affected bot processes and worked with the business to keep the business operations running until the outage was fixed.

In UiPath, DR can be planned with the help of the enterprise system. In ABC Insurance Corporation, there is an active-passive setup – that is, the active data center will host the entire UiPath platform, and all the products and components will be available in the passive setup. The passive environment will have a smaller number of machines and minimum resource configuration designed to support the vital business operations until the outage of the primary data center is fixed:

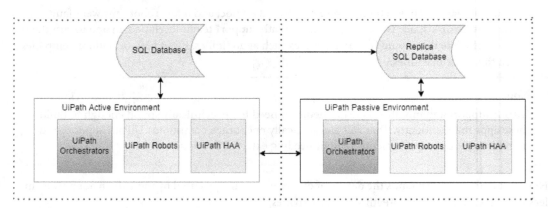

Figure 9.4 – UiPath disaster recovery active-passive setup

During DR, the UiPath support and administration team will be the primary team to work with all related stakeholders to keep the vital platform online so that it can run the business operations until the outages are fixed.

It is recommended to complete a mock exercise for DR scenarios at least once a year.

> **Note**
>
> Suppose the enterprise needs to have all the resources and machines available during the downtime of a data center. In that case, it is recommended to have multiple active data centers that host UiPath platforms. The cost of maintaining such a setup is a big consideration.

Now, let's try to understand the UiPath support team from an IT security standpoint.

IT security

The IT security team ensures all application teams follow an enterprise's IT security and data policies. This team's main objective is to reduce or eliminate data breaches and other security incidents.

UiPath developers need to ensure the IT security policies are followed by the bots that have been deployed in production. The UiPath support team also needs to be aware of IT security policies such as downloading files from external sites or emailing external IDs that contain sensitive business data. The UiPath support team needs to proactively escalate security issues to the UiPath development team to get them resolved using some internal security audit tasks.

In ABC Insurance Corporation, the UiPath support team contacts IT security for two main reasons:

- To get UiPath bot access for business applications and resources
- To help with the root cause analysis of an IT security incident involving a bot process

The UiPath support team needs to work on getting the event logs from application or database servers and application logs for the UiPath automation team to support the root cause analysis for the IT security team.

> **Note**
>
> UiPath bots are usually the primary suspects when unexpected business operations occur, such as locking multiple accounts or applications, making large API calls, and more. Hence, it is recommended that the UiPath support team understands the IT security policies in place to support such requests.

Now, let's learn how the UiPath support team handles requests from internal risk management teams.

Risk management

IT risk management deals with mitigating risk to business-critical systems in an enterprise. Internal audits are conducted to ensure business-critical applications follow the best practices and policies mandated by the IT governance team during development and support.

Since bots execute and interact with business applications and are vital to business operations, they are also in the scope of the risk management team. The UiPath support team will be contacted to help with an internal risk management audit of the UiPath bots in production that are deemed business-critical.

The application is usually categorized into different buckets, such as the following:

- Technology
- Business operations
- Regulatory

Ratings are usually provided based on the business criticality of the process, as follows:

- High
- Medium
- Low

This way, all the UiPath application stakeholders will know the most business-critical and high-risk application that needs to be supported compared to other applications in production.

In ABC Insurance Corporation, quarterly risk management audits are scheduled on UiPath bot operation, and reports are submitted to the IT leadership team on the findings.

> **Note**
> It is recommended to maintain a risk category for all the UiPath bots in a centralized location. The same information can be proactively shared with the risk management team during scheduled audits, reducing the UiPath support team's time to prove these reports on demand.

Now, let's understand how the UiPath support team handles requests from external auditing and compliance teams.

Auditing and compliance

IT audits and compliance checks are conducted using external auditors to ensure the IT governance policies and procedures are followed by all the IT teams, including UiPath CoE.

The auditors usually request different documents and production data to verify if all the standards are followed. In certain industries, compliance with certain state or federal rules must be audited. The UiPath support and administration teams will be requested user access details, application logs, and more. There might be several rounds of data gathering where the UiPath support team need support.

Audit recommendation usually has mandatory and recommended change items. Mandatory items need to be fixed within the reported time frame, while recommended items can be fixed when the team chooses to complete them. In both scenarios, the outcomes are recorded and shared with the audit governance team for future reference.

In ABC Insurance Corporation, external IT audits are conducted once a year. The UiPath support team will work with the external auditors to support the documents and logs of the requested UiPath application.

UiPath support leads will be summoned to discuss the audit report findings and to fix the gaps highlighted in the report. For instance, the audit report highlighted that 10 UiPath support documents contained outdated information and recommended updating the documentation as the support was outsourced to an external vendor. Once the 10 UiPath support documents were updated and reviewed, they were shared with the audit team to close this item.

> **Note**
> The UiPath support team needs to provide all the records, such as documents, to the audit team. Improvements are made in a centralized repository for future reference or questions.

Now that we have walked through all the advanced UiPath support topics with Jennifer, let's look at other UiPath product ecosystem support activities.

UiPath product ecosystem support

There are multiple products in the UiPath portfolio. ABC Insurance Corporation uses these five products shown in the following diagram. Jennifer needs to understand their capabilities to support them:

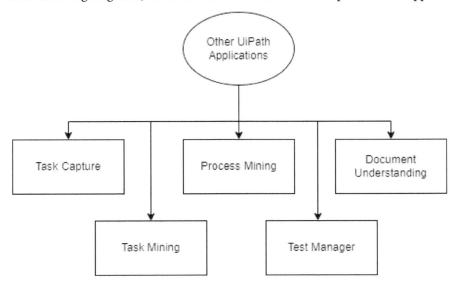

Figure 9.5 – Other UiPath products used in ABC Insurance Corporation

Let's look at these products in more detail and see how the UiPath support team is involved in handling some common request types.

Task Capture

Task Capture is a process mapping and documentation application that's used by business and IT stakeholders to document task-level information as flow charts:

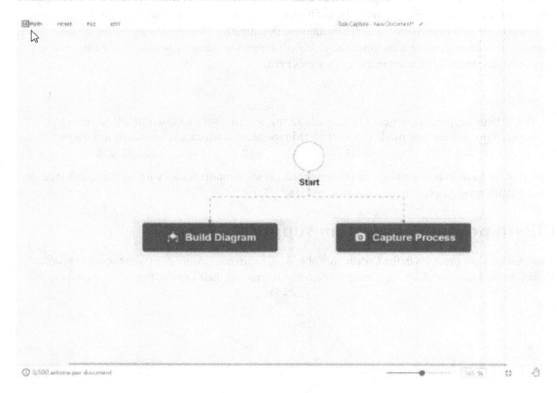

Figure 9.6 – UiPath Task Capture canvas

UiPath support team members will receive access to related issues from business stakeholders who will use this tool to capture process flows, enabling a prerequisites template to capture outputs or issues related to exporting output in different formats, such as words, images, and more.

Process Mining

Process Mining is a process analysis offering from UiPath that converts operational data from IT systems into interactive dashboards that can perform bottleneck analysis, understand critical paths, and improve the process before it is automated:

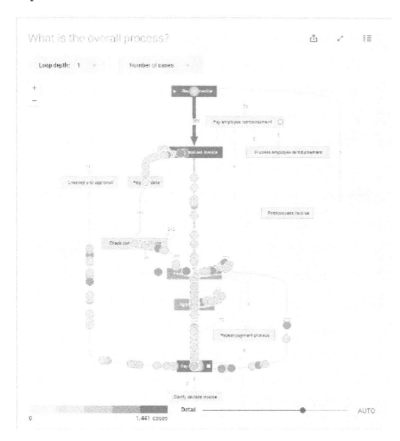

Figure 9.7 – UiPath Process Mining simulation

UiPath support team members will be mostly involved in setting up new data connections such as SQL Connectors, SAP connectors, or any custom data ingestion requests. They also support data refresh, archival, and performance issues.

Task Mining

Task Mining is an automated form of task discovery that works out of end user actions such as keystrokes and mouse clicks when they're performing a business task on their computer. The event data is then analyzed with a machine learning program that will be used to get real-time process data for process analysis and mapping or even finding automation opportunities:

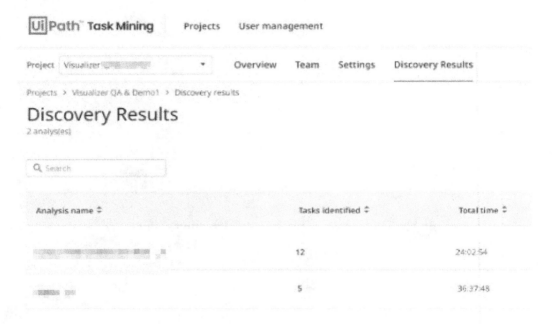

Figure 9.8 – UiPath Task Mining results

The UiPath support team will get involved with issues such as onboarding new users and uploading data from users' machines. They will work with data and business analyst teams to help them with filter and data visualization issues. Even working with developers and business analysts when they export to PDD or Studio can cause issues.

Test Manager

Test Manager is a web application that manages the automated test management process. Test Manager integrates with UiPath Orchestrator and Studio to link to automated test cases that have been created and available for use. This application can link, execute, and report automated tests:

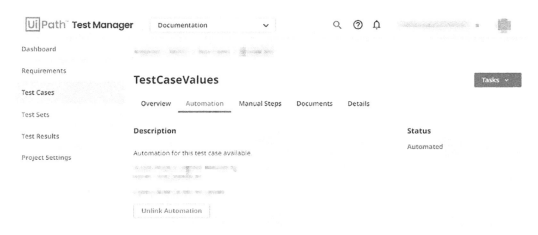

Figure 9.9 – UiPath Test Manager UI

The UiPath support team will be involved with requests from the UiPath development or test team to troubleshoot connection issues between Test Manager and Orchestrator. The support team will be contacted by the product management teams. If integrations with external systems such as Jira or ServiceNow are broken and need to be fixed. Testing and business analysts will reach out if the test reports haven't been updated or are not available in Test Manager. As this is a critical application, the UiPath support team must understand its capabilities to handle the support requests.

Document Understanding

Document Understanding is one of the most complex products that's widely used in the UiPath product portfolio to extract data from digital documents. Various operations are involved in the whole process; let's take a look:

- **Taxonomy definition**: The process starts with a taxonomy definition that explains the metadata of what data fields need to be extracted.

- **Digitize**: This step converts the document into a machine-readable format.

- **Classify**: The classification step will help you understand the category or type of incoming document.

- **Extract**: This is where the data from the rightly classified document will be extracted.

- **Validate**: If the confidence of the extraction is less, then the data will be sent to the human validation step. Otherwise, it can be sent to be exported directly.

- **Export**: In this step, the expected data extraction is reported to the requestor system:

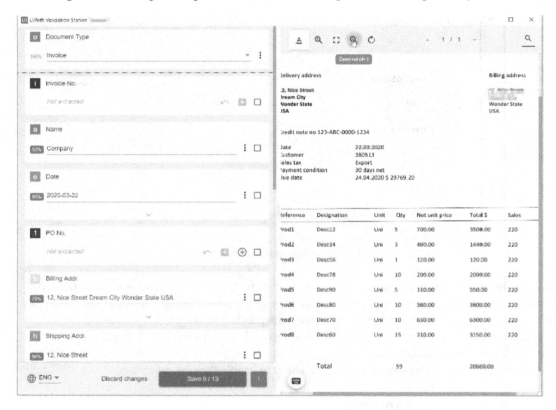

Figure 9.10 – UiPath Document Understanding – the Validation panel

The UiPath support team will be mostly involved if there are issues with the validate and export steps, where external operational team members are involved and unable to perform their actions. Few machine learning models are also involved in Document Understanding, which is one more area where issues may be raised concerning API keys or results not being returned from the ML APIs. As in many enterprises, huge volumes of data are extracted. This is a critical application where the UiPath support team may need product knowledge to complete the requests.

> **Note**
> We discussed some of the most used UiPath applications in this section, apart from the core ones (Studio, Orchestrator, Robots, and Insights), which we have already covered in this book. There are a few more UiPath applications, such as AI Center, Data Services, and others, available. It is recommended that UiPath support team members get an overview of these by visiting the UiPath product documentation website.

Now that Jennifer has covered the other UiPath platform support details, let's discuss the future trends that may interest the UiPath support team.

Future trends in UiPath support and administration

Technology disruptions were common trends in the recent past, and many trends potentially impact how UiPath support and maintenance activities are performed in UiPath Organization. Embracing technical innovations and adapting to change will benefit the UiPath support and maintenance teams. A few of the trends will be mentioned in this section.

Auto-healing bots

Most of the Level 1 types of UiPath support requests deal with restarting a failed UiPath job in production. In most situations, the fix will be simple enough – follow the SOP and resolve the issue.

Different technical solutions with advanced exception handling procedures will kick-start a cleanup job once the original job fails. This will act as a self-healing scenario where a faulted job will be fixed automatically without requiring manual intervention from the UiPath support team.

An ML-based ticketing solution

The UiPath support team mainly receives support requests from an ITSM-based ticketing solution, and in many organizations, the requests are still shared through emails or chats. As advancements in machine learning are happening every day, incoming support requests through informal channels such as email and chat can be categorized and assigned to the right support team.

The time that's spent understanding the requests from emails and creating tickets on the business requester's behalf will be reduced. This is a positive trend that will benefit the UiPath support and administration teams.

Container-based deployments and autoscaling

Docker is an operating-level virtualization platform that delivers software packages called containers. UiPath's official Docker images can be accessed by using docker pull: `registry.uipath.com/robot/runtime`.

We can use these images to spin different containers that can host UiPath robots in Linux machines. UiPath supports unattended automation to run Docker containers, which will allow you to scale the robots up and down at runtime based on the volume of the incoming request:

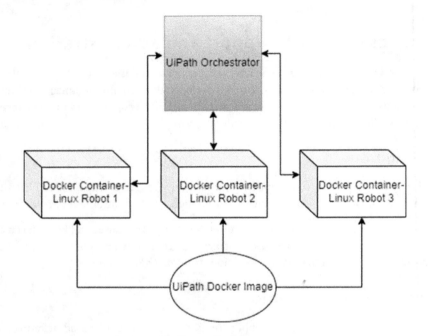

Figure 9.11 – UiPath Docker setup

These virtualization and scaling features will be part of the everyday UiPath support team; hence, having good knowledge of this technology trend will benefit the UiPath support team.

Multi-vendor ecosystem

The UiPath support team must be open to supporting multiple digital automation-related technologies apart from the core UiPath products. Different technologies such as workflow management, business process management, intelligent document processing, chatbots, machine learning apps, and more can be used along with traditional UiPath capabilities to provide an end-to-end solution:

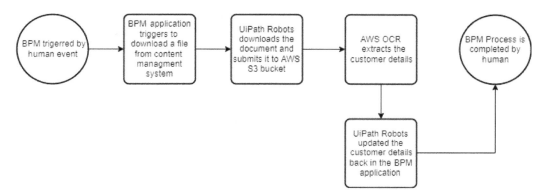

Figure 9.12 – UiPath multi-vendor workflow

In this scenario, the claims analyst in ABC Insurance Corporation used a BPM claims management system to request customer details from documents in the content management system. Once the request is submitted in the BPM-based system, a UiPath robot job is triggered, and the robot pulls the document from the content management system and submits it for data extraction in the S3 bucket. Once the extraction results are available, the UiPath robot sends the information back to the BPM system.

As the complexity of automation solutions increases, the UiPath support team members should be able to adapt to and learn the automation ecosystem to support production issues.

> **Note**
> Multiple RPA platforms such as Microsoft Power Automate, Automation Anywhere, and others can also be used in enterprises, along with the UiPath platform. Hence, having good knowledge of UiPath integration capabilities such as Orchestrator APIs and webhooks is necessary for UiPath support team members.

Automated support setup with advanced monitoring

Apart from the core UiPath analytics platform, UiPath Insights, there are many enterprise monitoring applications such as Splunk that can be invoked with the help of UiPath webhooks. This helps provide complete monitoring coverage from the application data and infrastructure levels.

Advanced monitoring and the alerting system will enable the UiPath support team to respond to issues before the disruption is felt on the business operations side. Hence, the UiPath support and monitoring teams must try to bring in the best-in-class monitoring solutions to enhance UiPath support and monitoring operations.

> **Note**
> Conversational AI enables end customers to interact with computer applications using text or voice as the interface. This is another important trend that will greatly impact UiPath support soon; hence, the UiPath support team needs to be prepared to benefit from this disruption.

With that, we have walked through all the topics of this final chapter with Jennifer. Now, let's summarize what we learned.

Summary

UiPath support and monitoring tasks become complex as the scope of RPA increases toward an integrated intelligent automation ecosystem. This chapter helped you equip new UiPath support personnel with knowledge about advancements in the field and give them an idea of the possible work items they need to prepare.

First, we looked at the different UiPath self-service support catalog items. This included the application, jobs, access, and monitoring services available to ABC Insurance Corporation UiPath stakeholders. The idea was to imprint the idea that you can enhance the customer experience of UiPath platform users.

In the next section, UiPath Orchestrator APIs were introduced, as well as different kinds of apps or utilities developed by the ABC Insurance Corporation UiPath support team. Again, the core idea was to help you understand how to automate the most mundane UiPath support and monitoring activities.

After that, we covered advanced topics such as BCP and DR, IT security, risk, and audits, where UiPath support team assistance is required. In the next section, a few popular UiPath products such as Task Capture, Document Understanding, and Test Manager were introduced. Common support requests for these applications were briefly explained.

Finally, we covered future trends such as container-based deployment, multi-vendor ecosystems, advanced monitoring, and more to help the UiPath support personnel understand what to expect in the future and to be prepared for it.

We walked Jennifer (from the ABC Insurance Corporation UiPath support team) through all the concepts in this chapter so that you can relate to that persona. I hope this chapter helped you get an overview of UiPath support advanced topics and future trends.

This brings us to the end of this book. I hope you have learned a few things related to UiPath support and monitoring from these nine chapters.

I hope this was a good use of your time! I wish you the best of luck in your UiPath support and monitoring role.

Index

W

Packt.com

Subscribe to our online digital library for full access to over 7,000 books and videos, as well as industry leading tools to help you plan your personal development and advance your career. For more information, please visit our website.

Why subscribe?

- Spend less time learning and more time coding with practical eBooks and Videos from over 4,000 industry professionals

- Improve your learning with Skill Plans built especially for you

- Get a free eBook or video every month

- Fully searchable for easy access to vital information

- Copy and paste, print, and bookmark content

Did you know that Packt offers eBook versions of every book published, with PDF and ePub files available? You can upgrade to the eBook version at packt.com and as a print book customer, you are entitled to a discount on the eBook copy. Get in touch with us at customercare@packtpub.com for more details.

At www.packt.com, you can also read a collection of free technical articles, sign up for a range of free newsletters, and receive exclusive discounts and offers on Packt books and eBooks.

Other Books You May Enjoy

If you enjoyed this book, you may be interested in these other books by Packt:

Robotic Process Automation Projects

Nandan Mullakara and Arun Kumar Asokan

ISBN: 978-1-83921-735-7

- Explore RPA principles, techniques, and tools using an example-driven approach
- Understand the basics of UiPath by building a helpdesk ticket generation system
- Automate read and write operations from Excel in a CRM system using UiPath
- Build an AI-based social media moderator platform using Google Cloud Vision API with UiPath
- Explore how to use Automation Anywhere by building a simple sales order processing system
- Build an automated employee emergency reporting system using Automation Anywhere
- Test your knowledge of building an automated workflow through fun exercises

SAP Intelligent RPA for Developers

Vishwas Madhuvarshi, Vijaya Kumar Ganugula

ISBN: 978-1-80107-919-8

- Understand RPA and the broad context that RPA operates in

- Explore the low-code, no-code, and pro-code capabilities offered by SAP Intelligent RPA 2.0

- Focus on bot development, testing, deployment, and configuration using SAP Intelligent RPA

- Get to grips with SAP Intelligent RPA 2.0 components and explore the product development roadmap

- Debug your project to identify the probable reasons for errors and remove existing and potential bugs

- Understand security within SAP Intelligent RPA, authorization, roles, and authentication

Packt is searching for authors like you

If you're interested in becoming an author for Packt, please visit authors.packtpub.com and apply today. We have worked with thousands of developers and tech professionals, just like you, to help them share their insight with the global tech community. You can make a general application, apply for a specific hot topic that we are recruiting an author for, or submit your own idea.

Share Your Thoughts

Now you've finished *UiPath Administration and Support Guide*, we'd love to hear your thoughts! Scan the QR code below to go straight to the Amazon review page for this book and share your feedback or leave a review on the site that you purchased it from.

https://packt.link/r/1-803-23908-5

Your review is important to us and the tech community and will help us make sure we're delivering excellent quality content.

www.ingramcontent.com/pod-product-compliance
Lightning Source LLC
Chambersburg PA
CBHW060524060326
40690CB00017B/3374